D0889644

**Management of
Engineering Projects**

Management of Engineering Projects

MAURICE SNOWDON
B.Sc., C.Eng., F.I.C.E., A.M.B.I.M.

NEWNES - BUTTERWORTHS
LONDON - BOSTON
Sydney - Wellington - Durban - Toronto

The Butterworth Group

United Kingdom	**Butterworth & Co (Publishers) Ltd** London: 88 Kingsway, WC2B 6AB
Australia	**Butterworths Pty Ltd** Sydney: 586 Pacific Highway, Chatswood, NSW 2067 Also at Melbourne, Brisbane, Adelaide and Perth
Canada	**Butterworth & Co (Canada) Ltd** Toronto: 2265 Midland Avenue, Scarborough, Ontario M1P 4S1
New Zealand	**Butterworths of New Zealand Ltd** Wellington: 26–28 Waring Taylor Street, 1
South Africa	**Butterworth & Co (South Africa) (Pty) Ltd** Durban: 152–154 Gale Street
USA	**Butterworth (Publishers) Inc** Boston: 19 Cummings Park, Woburn, Mass. 01801

First published in 1977

© Butterworth & Co (Publishers) Ltd, 1977

ISBN 0 408 00273 5

Typeset by Butterworths Litho Preparation Department

Printed in England by Chapel River Press, Andover, Hants

Preface

Everyone is interested in value for money. Value can be expressed in different ways such as the market price, the profit which can be derived, the pleasure obtained, security against risk etc. Whatever the criteria all will acknowledge the need to maximise the benefit from any cash outlay.

This book is concerned with the management and engineering aspects of this desire to maximise the value of capital investments. Engineering in its broadest sense is concerned with bricks and mortar, steel and concrete, pumps and pipes, wires and gadgets, paint and polish. The hardware is assembled in a particular manner, at a particular place, at a particular time in order to achieve certain objectives. This is what engineers have been doing for a long time, and always with a sharp eye open for the cash in the purse.

What then is new? In one sense there is nothing new under the sun but in another sense each assembly of hardware is new because it represents the creation of an asset which did not exist before. Each project is unique and presents a fresh challenge to skill, ingenuity and perseverance. The complexities are greater now than ever before. The pressures created by demands for speed, economy and superlatives of every description are such that more exacting demands are made of the creators of the assets than ever before. The achievement of maximum value in such circumstances is difficult but also important. It is a goal which must be constantly sought. It is a topic which amply repays careful study.

This book points some fingers at topics which are important and provides a background for the action. It will be of interest to those who contribute to the action from development through design to construction and operation. It will be of greater interest perhaps to those who control and manage these activities as well as to the client or developer who will be expected to pay the bills but who also reaps the benefits which his investment achieves. In the heat of design or flurry of construction it is always worth remembering that, whatever the pressures of the moment, the quality of the work in hand will be judged, day in and day out, by those who use or see the result of the project. The birth pangs will be forgotten but the product remains.

M.S.

Contents

Chapter 1

Setting the Scene

This book is about Project Management. It is written for Project Managers but above all for those who need Project Managers.

Unfortunately use of the term 'Project Management' does not immediately bring the same image to everyone's mind. It is therefore necessary to define the terms rather more closely.

Definition of Project Management

We shall be using the words 'Project Management' to denote the achievement of an objective by the creation of a new or the modification of an existing capital asset, i.e. a building, a plant or a facility. In considering the detail we shall be assuming that we are dealing with hardware and the design assembly and use of that hardware by people. We shall also assume that the management term includes the ability to choose, modify, design, shape, construct and operate that hardware to achieve a given objective.

No apology is made to those who use the term 'Project Management' for the launching and steering of an advertising campaign or the direction of a financial appeal for some worthy charity. Such use may be excusable, for the word 'project' has gained currency for any task that one might wish to achieve. The word 'management' has also become degraded. Every second person seems to qualify for the title so that it is sometimes divorced from the fundamental attributes of decision making and direction.

Two special cases merit attention. In some spheres, notably the contracting industry and certain types of manufacturing industry which have to set their stall out for specific and complicated tasks, the words 'Project Management' are used to identify the control of that 'project'. This stems mainly from the wider use of advanced techniques for

1

planning and cost control, both of which are of vital importance to profitability. A natural consequence is that Project Management has become synonymous with some of its techniques and a limited (albeit important) area of the whole activity.

It is the aim of this book to review such techniques but to place them in the context of a much wider project activity. Our concern is with the total project from concept through to commissioning. All the factors, issues, problems and techniques which are relevant to the project will be dealt with against this broader background.

The second special case is the use of the term Project Management to describe an organisational arrangement as distinct from a functional arrangement. This is dealt with more fully in Chapter 3 but we should note here the use of the title (see also 'Project Management, Swedish Style').

Project Management is an exciting sounding phrase. In the sense in which we are using it in this volume it is an exciting operation—full of the thrill of a creative action and the satisfaction of achievement. It is small wonder that the term is so frequently 'borrowed'.

In reality, Project Management, exciting though it is, is difficult, demanding and onerous. It is an activity with a considerable degree of responsibility and must not be under-estimated. It is a special branch of management and it will be seen through the pages of this book that we are constantly needing to define what Management is all about in order to deal with our subject. We shall also be discussing engineering so we might observe here that 'engineering' has come to mean something different as the term has been adopted to describe many of the craft trades. Some years ago a director of the Cessna Aircraft Company described an engineer as one who 'controls and makes useful to mankind things, processes and sources of natural and generated energy, the origin of which he does not understand completely'.

The definition has a sting in the tail but it is fair for all that and fits very well with another definition of an engineer as a 'contriver'.

It is only necessary to add that in so far as engineering is truly a profession, that project management is also a profession as this book will show.

The Objective

Before there can be a project there must be an objective which is worthy of attempting to achieve. Often the objective is ill-defined or becomes blurred so it is as well to stress that this question should be considered and the objective written down. It may have to be changed with the passage of time but the written objective remains a point of reference, as anchor against the storm.

It used to be said and still is that the objective of all industrial activity is to make a profit. Therefore the objective of an industrial project is to create assets which can be used for profit. It is more useful to see profits as a constraint on the man or company who would otherwise jog about unfettered and ultimately go out of business. That profits are necessary no one will deny but there will be other rather more specific objectives which can be expressed in marketing terms. It may be that a company seeks to double its share of a particular market by offering a better quality product at a competitive price. Or it may be that an entirely new product is to be introduced. Whatever the chosen objective it will have its constraints, of which making a profit will be one. We shall be considering the detail of these later but reference to *Figure 1* will be useful.

1. Technical knowledge
2. Licences and Patents
3. Available funds
4. Satisfactory profitability
5. Environmental considerations
6. Legal requirements
7. Available raw materials, services and labour
8. Market conditions

Figure 1. Typical industrial constraints

In the public sector the objective is likely to be expressed in different terms. The reason for building, for example, a library can scarcely be expressed in financial terms. The benefit in this case is in meeting a community need. It is not our function to comment on the reality of a particular need because that is a political judgment which will clearly depend on local circumstances.

It is however very much our function to say that the objective of any particular project shall be expressed in clear and unambiguous terms. This is true whether we are talking about the public sector or the private sector. During the course of the project many decisions will be taken and many compromises reached. The standard or reference point for any judgment which is necessary will be the objective of the project. Any decision which contributes to the achievement of the project must have a great deal of good in it and therefore this yardstick is a great comfort to those who have difficult decisions to make.

The NEDO Report 'The Public Client and the Construction Industries' published in 1975 pointed a firm finger at the trouble. 'There is often insufficient attention devoted to the client's brief' is one of the comments in the Summary and Recommendations. 'We recommend that the client body should always provide a clear and thorough project brief before design commences' (p.71).

Multiple Objectives

The objective of a project of course should be in the plural for it is not just a label to amplify the identity of a project. It should be a considered statement covering all the relevant aspects and must surely as a minimum run to several sentences. Questions to be considered in establishing objectives for an industrial project will include:

(a) *The scope*—What is to be made or achieved? In what form, size, shape, colour, specification? In what package? What materials are required to be incorporated? Are there any by-products or wastes?

(b) *The size*—How big should the unit be? How much product is required? Is there a choice to be made between one big plant and several smaller ones?

(c) *The place*—Is the location fixed? Where is the market? Where are the raw materials? Is there sufficient room for the project and any likely extensions? Is there sufficient labour for construction and operation? Are there sufficient services, e.g. water, gas, electricity? What transport is required? Is access satisfactory? Are the ground conditions satisfactory? Are there any environmental problems?

(d) *The money*—Has a realistic capital cost been estimated? Does it cover all the peripheral activities such as offices, warehouses and amenities? Has the cost of development work been assessed? Do the cost estimates include marketing costs? Do they recognise a period of time for the build up of output? Have operating and material costs been assessed? Are adequate funds available? Is the selling price structure established?

(e) *The timing*—Has a programme been committed to paper? Does it include all the prior stages? Does it allow for probable delays? Are the assumptions realistic? (some delivery periods are very long).

(f) *Will it work?*—Are there technical problems which imply a risk? Are there any novelties about the product or the process? Have reasonable assumptions been made with regard to breakdowns and unforeseen contingencies? Are any contingency plans necessary? Will it be safe?

These questions are not meant to be exclusive checklists. They merely serve to draw attention to the wide range of issues which must be faced during the life of an industrial project. They cannot all be tackled simultaneously but with knowledge of the specific project objectives can be established. To cover the gaps a priority list can be drawn up and outstanding problems tackled in an orderly and efficient manner.

The Ground to be Covered—Commercial Projects

In the case of commercial projects the objective will be to provide a service, e.g. a block of offices or a bingo hall. In this case the range of issues will be different from those listed above. Since there is a different meaning of the term 'product' some of the considerations assume a greater or lesser importance. It will however be perfectly obvious that the general message is equally applicable—there is a very large number of issues which needs to be brought into an integrated scheme of control.

The Ground to be Covered—Public Sector

For schemes in the public sector the same general point is also true but there might well be a greater concern with benefit rather than with profit. A particular problem in the public sector is the identification of the client. If a new town hall is to be built for example the claimants for prime interest include the rate payers, the council members themselves, the potential users of the facilities (i.e. organisations), the controlling department, central government etc. The headings listed above remain as issues to be dealt with (albeit with different emphases again) but there are even more groups of people to satisfy with the final outcome of the scheme.

In these circumstances the need for clarification of objectives and integration of control is even more vital.

One local authority had an almost trivial but troublesome example of the effect of not having sufficiently far-sighted objectives. A school was built in an area of new houses ostensibly for the children of that area. The nature of the school was then changed and the children were subsequently drawn from a much wider area. The consequence was that many children were delivered to and collected from the school in private cars. A large number of cars had not been foreseen and no special provision made. The result was a severe safety problem for a few minutes twice each day. The solution would have been to modify the school road and entrance systems but as an afterthought the expense was considerable. If this problem had been foreseen the extra cost initially would have been negligible for use could have been made of the delivery vehicle road system.

Of course it is easy to have hindsight and easy to criticise afterwards. The fact remains however that effort put into clarifying the objectives and writing them down is *always* effort well spent. Some of the problems will be forestalled.

The Eternal Triangle—Time, Cost, Quality

It will also be noted that the objectives have a time scale. Whether this is expressed in years, months or days will depend on the circumstances but it might be any of the three. There are many sad experiences of

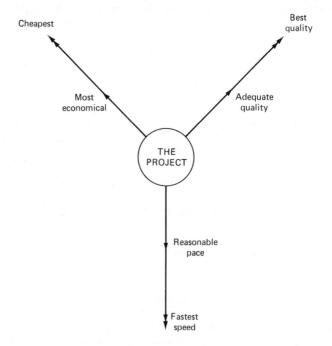

Figure 2. A delicate equilibrium

factories being built too late to meet the intended market. There are also cases of projects being abandoned part way through because the funds have run out and anticipated incomes have not materialised soon enough.

The idealist might demand that his project should be the fastest, the cheapest and the best that there has ever been. It is indeed praiseworthy to set high targets of achievement but it is equally foolhardy to suppose that such a formula is always or ever achievable.

The reality of the situation is that the vast majority of decisions required during the course of the project will be in the nature of having to decide which one of the three ideals will have to be sacrificed (*Figure 2*).

Take the simple case of deciding whether to work overtime in order to catch up on programme. Overtime usually demands the payment of premium time and if prolonged or excessive could also mean a fall off in productivity, i.e. even greater expense. Additionally, in some circumstances overtime or excessive overtime may prejudice the chances of securing the best quality.

Another common era where this conflict appears is in the provision of spare equipment. Should there be spares? How many? Should they be installed or available in the stores? The decision to have spares will cost money but could save time once the project is complete and might also improve the quality of the product or service being achieved.

Some of the more difficult decisions of this kind may well occur in the early project stages. If, for example, a development programme has been mounted to improve the quality of a product, how long can one afford to let such a programme escalate before deciding to proceed with production with the particular achievement which has been reached? How long will the market remain whilst the development work continues? Is the development work being done fast enough? Should more money be spent on development?

The solutions to all these problems will depend on very specific circumstances of the case. No general formula which fits all cases can be devised and it is certainly not the aim of this volume to stress one consideration rather than another. It is however important to recognise the conflict and to arrange that due consideration is given to all aspects whenever a decision has to be made.

The relationship between time and money is rather a special one which is often only partially understood. This will be referred to several times in later pages and in particular in Appendix 1. The relationship of these matters with the question of quality is more difficult to specify because it has a much more variable nature. In an industrial project it might influence the reliability of the plant being constructed or it might be reflected in the properties of the product, e.g. its colour or its shape or specification. In the public sector—on a road project for example—delay might perpetuate some dangerous traffic condition and time or money is then being balanced against human life.

Project Success

The importance which we have attached to a clear statement of objectives and the dilemma caused by the eternal triangle makes it imperative that we consider Project Success. If all the project objectives are achieved it is easy to conclude that the project is successful. But in fact this is only true if the objectives have been well written. This causes us to reflect that the criteria of success should be considered when the project objectives are being established and this is the reason for considering the issue at an early stage.

The criteria for success will probably include:

(a) Anticipated cash flow achieved.

(b) Designated benefit achieved.

(c) No unpredicted adverse consequences of a social, environmental, financial or commercial nature.

(d) No unwelcome impact on the sponsor's other operations or areas of interest.

(e) Completion to programme.

A project which is 100% successful by all relevant criteria is rare indeed but it is a goal that is well worth seeking.

Flexibility

Flexibility of project work must be emphasised from the start. In the pages which follow we shall be discussing many aspects and giving examples. No two projects are the same and even with the advantage of experience there is always the problem that circumstances may demand solutions different to those which gave previous success.

With this thought predominant this volume attempts to portray concepts and ideas rather than rigid disciplines. In the same way that projects themselves are flexible, the approach by management must also be flexible. An approach which is too rigid will have much less chance of success right from the beginning.

The Early Start

All who are successful in business will understand the importance of time. The significance of the earlier comments however is that blind abandonment to a race against the clock is not likely to be the optimum course. The message of these pages is the need for a balanced and controlled movement through all the phases of a project. This applies from the inception of the first idea through to full operational achievement. Balance and control imply planning and budgets (time and money again!) and a proper understanding of the quality requirements of the particular project under consideration.

Too often a project is not recognised as a project at all until it has gone through many of its initial phases, by when half the available calender time may well have been used up. Thus failure to achieve certain preliminary steps will result in the pre-empting of some choices which should have been available.

Getting the project under proper management at an early stage can only be to the advantage of the developer. This is the only way in which he can avoid being squeezed at later stages. It is the only way in which he can be sure that he has the best chance of success. It is the only way in which he can be *sure* that adequate effort is being deployed on all the issues which affect the project under consideration.

The need for adequate management of a building project in the early stages is reflected in the HMSO booklet 'Before you Build'. Having quoted that 20 000 or so clients each year obtain industrial or commercial buildings it reports that 20% were dissatisfied with the service they received from the construction industry. Reading that 25% of the clients had not clearly established their requirements before they approached the construction industry, their dissatisfaction is not surprising. Indeed we should be thankful that the construction industry seems so often to have clarified the client's requirements to the client's satisfaction!

Since it is dealing primarily with buildings and refers almost exclusively to architectural and quantity surveyor roles the booklet is only in part applicable to the situation which this volume envisages but the message holds true. Satisfaction at the end of the day can only come from adequate preparation. A project is often a 'one-off' job and usually expensive so it is patently foolish to run the risk of dissatisfaction by inadequate or late preparations.

Pitfalls Galore

During the life of a project many decisions have to be taken and many events will arise which will shape the results. As far as possible a positive approach is preferred, steering a course for home between the hazards which can be foreseen. These will include the following.

Disaster

The incidence of the 1974 Flixborough disaster brought vividly to mind the possibility of major fire, explosion or major accident. This possibility

" must not be overlooked and due regard must be paid to the materials and processes which may be found in the plant. Safety is dealt with in more detail in Chapter 2, but the point here is that an accident may not only be expensive in itself and involve human suffering but also its consequences may be far reaching. The loss of a key component could mean a loss of production lasting months and which could be crucial in terms of cash flow.

Of course insurance plays an important part in all this but insurance premiums are heavy and increasing. The wise developer will wish to cover himself adequately with insurance but also to ensure as far as possible that his project is well conceived and not outstandingly dangerous or accident prone.

The public sector is not immune to disaster as regrettably serious incidents have shown. The outcry against high alumina cement and the caution associated with the Merrison report on box girders serve as examples of areas which have caused deep concern. The risk of disaster is real; but it needs to be faced with cool judgement with a keen and analytical search for likely trouble spots.

Cash Flow

The financier may regret that in order to convert his capital into a production unit he has to go through the many traumatic experiences that a project usually brings. He may castigate those whom he thinks responsible for delays, mistakes or over-expenditure, especially if he fails to understand the climate in which the whole activity is set. There is also the question concerning the availability of enough cash to permit the plans to be implemented.

It should be recognised by all concerned that estimates of time and money cannot be exact. It is indeed a very good capital estimate which is right to within 10% and in inflationary days it is surprising that any estimate bears comparison with the ultimately achieved costs.

The same problem applies to programme dates, although programmes tend to be even less accurate than financial estimates—usually because data is not collected in the same way. It is particularly true if technical uncertainties are involved.

The combined effect of these difficulties is that the required cash flow is very hard to predict. It is therefore necessary to ensure, before starting a project, that the cash flow predictions have a sufficiently large margin to meet all contingencies. There is no joy or credit in a partially completed project which has had to stop because of lack of funds. Abandoned plant and equipment deteriorates rapidly and has a very low market value.

Absence of Market

This is perhaps the most heart-rending pitfall of all. If the product will not sell—whether it be overpriced, below specification, outpaced by competition or just not wanted—there is no income to set against all the charges that have been incurred. In some projects there will inevitably be a high risk of this happening. For such projects, if successful, a correspondingly high return will be expected.

Projects which are prone to this hazard should include two provisions:

(a) A programme which delays the bulk of the spending until the latest possible time, and the incorporation of a review mechanism at the latest possible time before spending decisions are finally taken.

(b) A critical review of the circumstances in which the market could be lost so that all facets of the problem can be understood and evaluated. This review would normally be undertaken early in the life of the project so that alternative courses of action (escape routes!) can be identified. It would be appropriate to update the review during the life of the project at reasonable intervals.

Lateness

This has been mentioned under 'The Early Start' and more will be said in Chapter 5. The importance of early planning cannot be too strongly emphasised and the absence of adequate forethought remains one of the most serious of all the pitfalls.

Effect on Environment

In the majority of cases this is predictable and is frequently settled at the planning permission stage. Certainly this is true of visual amenity. The appearance of a building is capable of demonstration by a drawing and even if the effect turns out to be less satisfactory than expected it is unlikely that there will be any argument on the facts of the situation.

There is more room for opinion on problems of smell and even this will vary according to the wind. Similarly emissions of fume, smoke or effluent may be less difficult to predict with complete accuracy although these have more chance of being measured on a scale of numbers.

Noise is also very much a conservationist's hobby-horse at the present time. This is not to decry the importance of adequate consideration to the problem but it should be removed from emotive arguments. Noise can be reduced although this is sometimes expensive.

Given adequate and early consideration, environmental problems generally should not be a pitfall. They are capable of prediction and largely can be measured. They are included in this list of pitfalls because public opinion now demands adequate sureties.

Transport Problems

These should be included in the list because they may be wider than at first supposed. No doubt the particular requirements for delivery of raw materials, removal of product and by-products, relative merits of road, rail or water transport, access for fire and service vehicles, parking of staff cars, etc, will all be considered at the right time. As a result of this consideration the internal factory transport system and entrances will be laid out.

The example of the school transport problem quoted earlier demonstrates that this is equally a problem in the public sector. We may be borrowing from the realm of science fiction if we think that all buildings should have heli-decks but perhaps we wouldn't even have had the thought when most of our public buildings were erected!

The wider issue may be the route of road vehicles away from the proposed site. This is partially an extension of the environmental problems for transport tends to create a noise nuisance particularly if it operates at night. Increasing weight of road transport units also brings vibration issues to the fore as well as accident figures, heavy vehicle parking problems, traffic congestion and such matters.

These issues may not actually prevent production or use from taking place but will cause the development to be viewed in these senses as antisocial and unpopular. Again it is a pitfall which can probably be avoided by early consideration.

Wasted Rescources

The project will consume resources in the form of individual effort, energy and raw materials. This applies to all stages of development, design, construction and operation. The wastage of resources is uneconomic rather than preventing absolutely the completion of the project. The effect will therefore be felt in the profitability of the venture which is one of the more painful ways of feeling the effect.

Some 'waste' is inevitable because no processes—human or other-wise—are 100% efficient. The task therefore should be seen as an optimisation of efficiencies. To do this properly is amply rewarding. It means drawing up balances for materials used and for energy as well as detailed assessments of the labour requirements for all phases of the job.

The pitfall in this case is not so much a deep well as a morass of very sticky mud!

The Law Breaker

With the increasing complexity of legal requirements it becomes more and more difficult to avoid breaking the law inadvertently.

For any particular industry or type of project it is possible to draw up a list of relevant legislation so that most of the guide lines can be understood. At the commencement of any major project it is worth while doing exactly this by ensuring adequate consultation with local authorities, factory inspectorate, fire authority, the appropriate government departments, etc.

The Dusty Shelf

No prospective project manager or developer likes to think of his uncompleted scheme being consigned to the archives or the filing cabinet or (worst indignity of all) the W.P.B. It is a fate which we hope does not come your way.

Nevertheless it is far better to abandon a scheme than to pursue it recklessly to an unprofitable conclusion. Realistic and dispassionate assessments carried out in good time will cut losses and save undue embarrassment from any scheme which is destined for abandonment.

Chapter 2

The Anatomy of a Project

The anatomy of a typical project is shown in *Figure 3* and more simply in *Figure 4*. Since no two projects are the same, variations can be expected and in particular some recycling between the overall financial assessment and the earlier stages is common. This is especially true of a high risk high capital job which will need very detailed and careful analysis before financial approval is given.

The individual steps in *Figure 3* demand a closer look.

Need for Capital Development

The start of the whole process of project work lies at the point when the need for facilities has been identified (or presumed). In the industrial scene this will have been brought about by expectation of a sales potential or a service requirement. There are several possible triggers such as:

Make a new product.
Make an improved product.
Make more of an existing product.
Make an existing product more economically or acceptably.
Exploit a new market, etc.

In the case of the public sector someone will have identified a benefit to be achieved. This could be a bus station or a hospital, a reservoir or a road. Most projects in the public sector are triggered by an expression of public need whether this is initiated by a committee of elected representatives or officers of the public body undertaking their responsibilities.

15

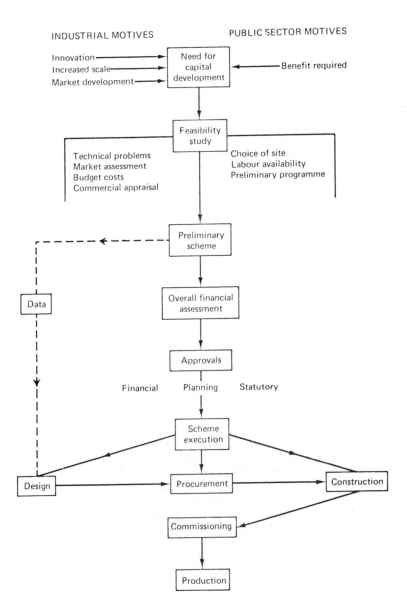

Figure 3. Project steps

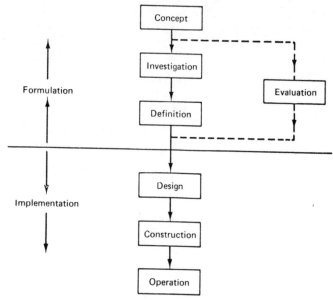

Figure 4. Simplified project anatomy

Maybe the trigger will be just an idea or a tentative suggestion. It may only be a hope but, however flimsy the vision may be, it is as well to recognise that the potential developer has got a project on his hands even at these very early stages.

Feasibility Study

Whilst the main motive may be clear enough the mere desire to meet a chosen objective does not necessarily mean that it is the right thing to do. It is likely that the first concept needs modification in order to improve the chances of it being successful. In any case, as we have seen in Chapter 1, the project objectives will need to be expanded and more detailed. Whether the action at this stage is formalised or not there will in fact be a feasibility study carried out.

This will attempt to forecast how the scheme might be carried out and test each aspect to see if there are any particular problems or unforeseen difficulties. At this stage a rapid survey using a comparatively small resource is required.

A feasibility study is like standing back and taking hard cold look at the proposals. Often the enthusiasm for the project masks some of the snares which will later prove expensive and cause delays.

It is also necessary that any technical risks involved should be put into their correct perspective. Risks are normally a recognised part of a development programme and the project which follows but they always need to be quantified and assessed.

A feasibility study would normally expose all the facts of a situation and demonstrate areas of uncertainty. It would consider the problems associated with the proposed development from the several points of view of engineering, performance, economics, environment, programming and cash flow.

Specific investigations which may need to be undertaken or planned include visual inspection of the proposed site or sites (see *Figure 5*),

MAIN ISSUES

1. Availability of raw materials.

2. Disposal of products, by-products and waste.

3. Available area.

4. Availability of labour – skilled and unskilled.

5. Availability of services – especially water.

6. Impact on environment.

7. Special requirements of particular scheme – if any.

SECONDARY ISSUES

1. Requirements for future extension.

2. Ownership or present use considerations.

3. Problems of planning permission.

4. Nature of ground – drainage and loading.

5. Site contours.

6. Underground features – mining etc.

7. Proximity to airfield (height considerations).

8. Access required – lorries, cars, rail, other.

9. Ancillary facilities required for labour in construction and operation, e.g. housing, recreation needs, community matters.

10. Site ancillaries and their impact, e.g. canteen, laboratory, warehouse, workshops, offices etc.

Figure 5. Site selection check list

18

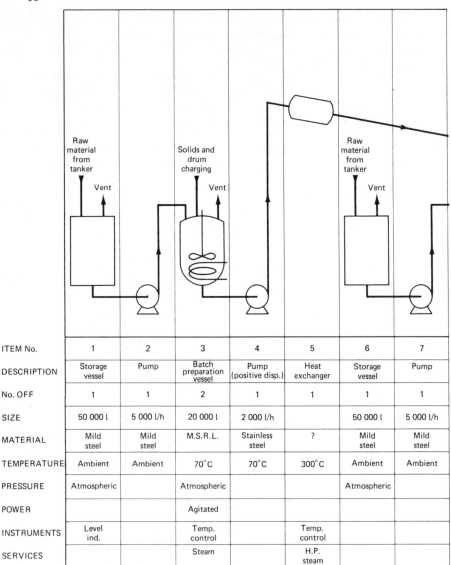

ITEM No.	1	2	3	4	5	6	7
DESCRIPTION	Storage vessel	Pump	Batch preparation vessel	Pump (positive disp.)	Heat exchanger	Storage vessel	Pump
No. OFF	1	1	2	1	1	1	1
SIZE	50 000 l	5 000 l/h	20 000 l	2 000 l/h		50 000 l	5 000 l/h
MATERIAL	Mild steel	Mild steel	M.S.R.L.	Stainless steel	?	Mild steel	Mild steel
TEMPERATURE	Ambient	Ambient	70°C	70°C	300°C	Ambient	Ambient
PRESSURE	Atmospheric		Atmospheric			Atmospheric	
POWER			Agitated				
INSTRUMENTS	Level ind.		Temp. control		Temp. control		
SERVICES			Steam		H.P. steam		

Figure 6. Typical flowsheet

Notes: 1. The process is entirely mythical—the format is important.
3. Routine information, e.g. valves, is kept to a minimum to
focus attention on the fundamentals

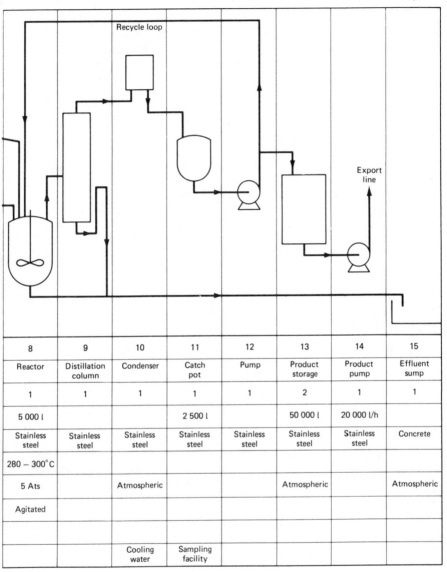

8	9	10	11	12	13	14	15
Reactor	Distillation column	Condenser	Catch pot	Pump	Product storage	Product pump	Effluent sump
1	1	1	1	1	2	1	1
5 000 l			2 500 l		50 000 l	20 000 l/h	
Stainless steel	Stainless steel	Stainless steel	Stainless steel	Stainless steel	Stainless steel	Stainless steel	Concrete
280 – 300°C							
5 Ats		Atmospheric			Atmospheric		Atmospheric
Agitated							
		Cooling water	Sampling facility				

for a chemical process

2. Gaps in the table indicate unknowns–probably numerous
when the flowsheet is first prepared.

subsoil investigations, comparison with local authorities' plans and requirements, availability of labour for construction and operation, technical aspects specific to the proposal, etc. Market studies can also form part of the feasibility study if appropriate.

Speed is of the essence and feasibility studies will always be executed quickly. The aim is not to do design work or to solve problems but to indicate the plan of attack which is necessary to achieve the objectives. The cost of executing a feasibility study is not usually therefore very great and is always small in relation to the benefits accruing.

Whether the study is being conducted by the developer using his own resources or whether an outside group is being employed, the requirements from the study and the terms of reference need to be agreed at the start. It is a matter of experience that some ill-framed feasibility studies never get completed and get hopelessly mixed up with later stages of the activity. Good direction and proper management will save confusion.

Preliminary Scheme

Almost as part of the feasibility study a preliminary scheme needs to be drawn up so that costs can be identified. In the chemical or petro-chemical industry this would be a flowsheet but the term does not have the same significance in other types of project (*Figures* 6 and 7).

To achieve this stage the proposals must be thoroughly identified in order to:

(a) Ensure that all the steps necessary to achieve the objectives have been defined. This not only means that the objectives will have been defined sufficiently to enable the full scope to be recognised but also that a lot of the detail has been considered at least superficially to foresee potential problem areas.

(b) Be able to identify a reasonably accurate capital cost of all the items which will be required.

(c) Be able to estimate the service requirements, e.g. water, gas, electricity, steam, etc, with reasonable accuracy.

(d) Establish labour requirements for the steps envisaged.

(e) Be certain, in the case of projects which require them, that all the raw materials or components can be listed and their availability assessed.

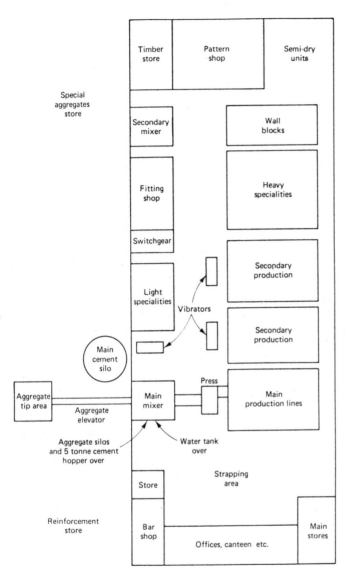

Figure 7. Preliminary scheme for an industrial project
Notes: 1. *The project is for pre-cast concrete manufacture.*
2. *The area of land required is four times the area of building.*
3. *Road access for heavy vehicles is essential.*
4. *Services (water and electricity) are required*

The end point of this stage is usually determined by the achievement of definition described by:

(a) The nature of the product or the benefit and its specification (especially the limiting condition, i.e. what the project will *not* achieve).

(b) The level (or levels) of production or operation identified as targets.

(c) The manner of achievement and the major steps to be achieved recognised.

(d) All characteristics of prices and costs and their variabilities recognised (capital, operating, maintenance, materials costs and all revenues).

(e) The site or sites of all the activities chosen.

(f) Action required on environmental matters, effluent streams, by-products, etc, defined.

With the formulation of the Preliminary Scheme (assuming it is satisfactory) we have really come to the end of the pre-design period. The next two steps are not usually contributors to the technical requirements of the project but until they have been achieved it is premature to open up the high volume of activity which lies within scheme execution. This is therefore something of a twilight zone but it can be put to good use by the formulation of Project Data. We will discuss this in more detail in Chapter 3 (and Appendices 3 and 4) but should note here the important link between the Preliminary Scheme and the Scheme Execution. If forged well this link will achieve two functions.

(i) It will allow the design work to get off to a flying start at the right time.

(ii) It will maintain the continuity of the purpose of the project through the crystallisation of the objectives.

Overall Financial Assessment

From the preliminary scheme an assessment can be made to determine:

(a) whether the scheme is sufficiently profitable or attractive to proceed.

(b) how much capital will be required and when it must be available.

It is not the purpose of this volume to consider what represents a satisfactory return on capital or a satisfactory benefit. Even in the private sector this will vary between industries and between products within those industries. It will also vary with the risk which is being undertaken. Clearly the company which takes a risk in investment will expect a higher yield than the company which invests in certainties.

In the public sector it is even more difficult to lay down criteria of acceptability. If a decision rests solely on environmental improvement or the provision of a service it is likely to have a considerable local significance and the political logic may stem from local or national level.

Whatever the project we can insist that the financial implications should be set down and studied. Social benefit is difficult to assess in financial terms but money is a common denominator in our civilisation and we should recognise the cost even if we cannot agree on the value.

The assessment however will be an assembly of forecasts or predictions and not a statement of facts. Since engineering projects typically take one, two or more years to complete and hopefully have an operational life which is much longer the problems of forecasting with accuracy are very real. Each element must be taken on its merits and may be treated either as a sum of money with limits of confidence or as a range of possible costs.

Just to take one example from the public sector think of the difficulty of predicting the revenue from a new swimming bath. This will be the sum of all the products derived by multiplying the number of patrons by the prices of admission (ignoring private hirings or miscellaneous other sources of income). For any one period there are likely to be several prices of admission (e.g. swimmers or spectators; children, adults or pensioners; afternoons or evenings; weekends or weekdays, etc.) All these prices can be governed by choice (whose choice?) so if the number of patrons could be fixed the revenue could be estimated.

But the number of patrons cannot be fixed. It will depend on many factors such as: admission price (!), quality of facilities, catchment area, population habits, education authority attitudes, weather conditions and rival attractions.

The best one can hope for is to establish some probabilities such as an absolute maximum, a probable absolute minimum, and some view as to the most likely range in practice.

But the revenue from the project is only one of the elements in the assessment. We have also to consider the capital cost and the interest on the loan charges, the operating cost, the maintenance bills, the life of the asset and perhaps other elements such as rent from the restaurant operators, etc. Each will have uncertainties which in some way will have to be evaluated if a meaningful assessment is to be made.

We have ignored inflation which of itself might be the biggest question mark of all. Our purpose has been to show that we have the problem of combining a number of uncertainties together in such a way that we have a result which is at least good enough as the basis for an important decision, i.e. spend or not.

In a practical case this problem must be considered in detail and the bibliography may be consulted for further study. We must however give some attention to the technique known as Risk Analysis and some further notes with the industrial scene in mind are included in Appendix 2.

We shall also note that there are several techniques available to compare projects once a prediction begins to emerge. If you have a sum of money available you will want to know whether you will gain more by leaving it in the bank or authorising the project. But you may also want to know whether you should spend your limited money on Project A or Project B. Thus not only does the financial assessment have to be as accurate as possible in absolute terms but also in comparison with other contenders for resources. However it is not the purpose of this volume to advocate one method rather than another, though we must mention that the choice may include such methods as:

Interest rate of return (see Appendix 1)
Equivalent annual cost
Simple return on capital
Payback period, etc.

The purpose of this particular stage is to establish whether the investment should be undertaken or not. If it should we proceed down the chart. If it should not we may either abandon the project completely or return to an earlier stage to re-assess. We shall assume that the decision to proceed is taken.

Approvals

Before the rate of expenditure on the project becomes too large it is necessary to ensure that all the required approvals have been obtained or alternatively the absence of such approval is recognised and acceptable. This latter situation might imply some risk if a required approval should eventually be withheld but this may be acceptable if the risk involved can be satisfactorily evaluated.

If it is a private company the particular organisation or the sponsor seeking to execute the project may demand that he should gain approval or support from parent boards, sales control groups, etc. These internal arrangements will be taken for granted here. If the sponsor is a public body it may be necessary to gain approval from central government or other affected parties. Again such organisational situations will be assumed to present no problems although in reality this is often far from the case.

The nature of the necessary approvals at this stage will therefore be:

(a) *Financial*—In preparing the financial assessment all the available financial data will have been incorporated into a prepared statement. This will be aimed at satisfying the body which holds the funds. This might simply be the owner, the board of directors or the public body charged with this particular responsibility.

The criteria to be judged will include:

(i)   Are the assumptions in the assessment as accurate as possible in the light of all circumstances?

(ii)   Are adequate funds available at the right time to execute the project?

(iii) Are the risks involved acceptable and justified in the attempted execution of the project?

It is wrong and damaging to withhold or delay approval without good cause. Whatever flaws or uncertainties may appear to the approval body they must be recognised and dealt with so that the consequences, whatever solution is chosen, are known to be acceptable. The stage of financial approval is a time for coolness, calculation and courage.

(b) *Planning*—The feasibility study should have shown what problems might be encountered over planning permission but this will be the time to actually obtain the appropriate approval. If detailed drawings are not available (and they probably won't be!) the application may have to be for outline planning permission in the first instance. In any case an initial informal approach to the planning authority is recommended because it may well be that, in the case of a significant project, the authority will require additional information which can best be identified in discussion. This might well include assessments of environmental impact, numbers to be employed etc.

Environmental Impact Assessment is now something of a jargon phrase which will either become increasingly important or be superseded by more acceptable techniques. In essence it is an attempt to quantify effects which hitherto have been largely subjective. Whatever benefits a project may bring either in cash reward or social advantage it is likely that there will be at least some associated disadvantages, even if experienced by only a small minority.

For example a new road may be judged to benefit motorists, industry and commerce. It may relieve unwanted congestion somewhere and it may increase safety. Generally the social and financial benefits may be highly regarded but there may still be loss of agricultural land, uprooting of trees, demolition of buildings and risks of noise or visual intrusion. It is clearly difficult to balance such factors because they are dissimilar in nature. Public outcry against some of the more controversial proposals have led to fresh attempts to make comparisons more meaningful, and E.I.A.s are one result.

The increasing awareness of the importance of conservation has meant greater pressure to undertake assessments. To be of real value they have to be comprehensive and therefore expensive and time consuming. There is therefore considerable scope for short cut techniques which will deal more satisfactorily with a wider range of projects.

(c) *Statutory*—Certain types of development require additional statutory approval. Since these tend to change with political whims it is difficult to give a list which can be regarded with certainty. Some of the more obvious and less political ones might include:

> Height of buildings near flight paths.
> Cross-country pipelines.
> Petroleum installations.
> Use of asbestos.
> Interruption of water-courses, etc.

Scheme Execution

This is the stage at which activity begins to expand dramatically. Many people are becoming involved and the project appraisal is likely to underline the need for speed. This emphasises the importance of having done the necessary pre-planning and scheme work to sufficiently high standards. Typically 3–5% of the total project cost will have been spent in development work, feasibility study, assessments and prior stages. Nevertheless 80% of the total cost will have been determined unalterably. This is the point at which the spending of the remaining 95–97% begins in earnest.

For many people the scheme only consists of one, two or three of these parts, i.e. Design, Procurement and Construction. The earlier parts must have been achieved for them to do their work but they will not have been involved in the earlier stages. This induction problem coincides with acceleration of the activity. The work to be done in this phase is described in detail in later chapters.

Commissioning

The commissioning stage is deceptive to the uninitiated. If start-up were as simple as pressing a button or a switch everyone would be delighted but this is rarely the case. To examine this stage in some detail the following subdivisions may be discerned (not necessarily in strict chronological order).

(a) Verification that the work has been completed to the approved design. This is an overlap of the construction activity and needs to be done once thoroughly and not twice partially.

(b) Performance testing to acceptable standards. In the case of a pump for example this may already have a supplier's test certificate but maybe with a different motor and maybe the seals have been changed. The need for further tests will depend in part on the importance of the performance in the context of the whole operation.

Other items, such as a conveyor, may have been assembled on site and some testing is warranted before the item is exposed to its permanent function. This is also the right time to see if furnaces heat up as quickly as expected.

There are classic faults to be avoided such as fans wired up to rotate in the wrong direction, blanks left in pipe lines, overheating of bearings, etc. Methodical verification is required and should achieve a high degree of confidence.

(c) Dummy runs (especially in an industrial project). If the material to be handled is expensive or hazardous in any way it is common practice to arrange some dummy runs or 'water trials', the object generally being to ensure that consecutive pieces of equipment do in fact fit together in the required manner. Some instruments can be proved in this type of test.

(d) Operator training may already have begun during construction but it will be remembered that the end of the commissioning stage is routine operation. The operators will require familiarity, precise instructions and some experience for them to be able to achieve their tasks. In detail this will vary a great deal according to the complexity of the project but the initial success of the venture depends greatly on this aspect.

(e) Stage-wise operation may be necessary in a multistage project or process. It may also be necessary to try out the latter processes first to pave the way for subsequent successful operation of the earlier stages. Stage-wise activity may also be necessary to achieve quality checks at intermediate places in production lines.

(f) Planning and programming of all these activities is obviously necessary particularly as the pressures will be very severe with the majority of the capital spent and a great anxiety to achieve some return with all possible speed.

Art or Science?

Having dissected a project and analysed its components to a degree one would be tempted to conclude that project management is a science. Alternatively one may think that the question is irrelevant and that it doesn't really matter whether it is or not. In fact project management is a blend of both art and science.

The numerate side is obvious. Here we are concerned with time and cash and both of these are measured on well recognised scales. The constant need for a rational approach is a reminder of the scientific method.

Many projects are industrial, and based on a scientific process. Whether this be stitching shoes, making concrete or cooking chickens does not alter the fact that the process follows scientific principles. Instruments and control systems will be used to achieve a defined process within specified numerical limitations.

Similarly, in the public sector, that which is being created depends very largely on scientific principles for its function and for its creation. Nevertheless the point of raising the question at this stage is to create the right balance between a strict scientific approach and what can best be described as an art.

Engineering used to be an art although it is less so now. In a project there is room for flair and imagination—there is room for an individual to apply his talents. We have noted already that there is bound to be backtracking and adjustments. This is to allow for changes, and the project manager must become accustomed to the problems of managing change. This is where his skill comes in for he must be no mere automaton making decisions which a computer could make if the correct data and the correct programme were available.

Each project is different, so it is impossible to use a carbon copy of one job to make another. This also is why individual decisions have to be made by people and the management function cannot be programmed.

Furthermore the successful execution of a project will depend on people contributing towards the achievement of the whole. The management of people, in spite of behavioural scientists, is never going to be completely scientific in its outlook or implementation.

We shall return to the human side again in Chapter 6, but the point at the moment is to emphasise that although we are examining the components of a project in a logical and scientific manner we must not neglect the fact that there is this other facet to the problem.

Safety

Recent years have seen an upsurge in the attention given to health and safety as measured by the weight of legislation or by the activities of the media. Whilst such attention is welcome there are undoubtedly digressions which, however well-meaning, tend to distort the balanced view which is essential in managerial decisions.

This, in itself, justifies devoting some thought to what is really a very complex subject.

The engineering profession has always had to operate near to the boundaries of knowledge without necessarily foreseeing danger. Every now and again the problem surfaces in the report of a major disaster—the Tay bridge collapse, dam failures, the sinking of the Titanic, air or train crashes, the Flixborough incident and so on. Such public denouncement these incidents receive merely serves to re-focus attention on the age-old problem. The Romans built aqueducts which failed and any attempt to make progress involves the possibility that some important feature has been overlooked and risk is involved.

It is not surprising therefore that the construction industry at large has an unflattering safety record. However safe the ultimate achievement of a project may be there is always a high-risk period to be passed through.

On the industrial front Great Britain has a record which might be judged better than other countries. Nevertheless in absolute terms there is no doubt that industrial accidents, even in Great Britain, are responsible every year for many deaths, a great deal of suffering and an enormous amount of lost time. Each year there are something like half a million reported accidents. Also, about 20 million days work are lost each year as a result of accidents.

Clearly there is room for improvement whether one's motives are moral, humanitarian or economic. One can but approve of every step which seeks to improve designs and reduce unknown factors to a minimum. The vast increase in British Standard specifications and Codes of Practice bears testimony to the fact that a great deal of knowledge is being assembled and made available for the benefit of all.

And yet the problem remains! In spite of such accumulations of knowledge, safety still depends very largely on the human element. The majority of accidents occur either because a person has committed an error or because someone has failed to perform some duty or follow some instruction.

The acknowledgement of the human element in safety is not new but a fresh approach to it was used in the preparation of the Robens Report in 1972. Many aspects of the report are worth quoting and selection is regrettably discriminating. Nevertheless to encourage a wider reading we quote:

> Equally important factors in safety and health at work are the attitudes, capacities and performance of people and the efficiency of the organisational systems within which they work. (Page 8)

> The primary responsibility for doing something about the present levels of occupational accidents and disease lies with those who create the risks and those who work with them. (Page 7)

> The most fundamental conclusion to which our investigations have led us is this. There are severe practical limits on the extent to which progressively better standards of safety and health at work can be brought about through negative regulation by external agencies. We need a more effectively self-regulating system. (Page 12)

Following the Robens Report there has been the Health and Safety at Work Act 1974 which has been justly greeted as one of the most fundamental pieces of legislation since the Factories Act. Very significant changes were introduced in the Act and it signalled alterations to the whole structure of the Factory Inspectorate. The introduction of prohibition and improvement notices together with stiff penalties for non-compliance in certain circumstances has sharpened the enforcement of the legal aspects. But this is still only half of the problem.

The other half of the problem, recognised by the 1974 Act but not capable of being dealt with by legislation, is that safety is largely a function of awareness. No matter what equipment or design or instruction is devised many accidents are caused by thoughtlessness or human error, or by a combination of unforeseen circumstances. 'Safety awareness', says Robens, 'must be deliberately fostered'. 'The most important single reason for accidents at work is apathy. ... safety is mainly a matter of the day-to-day attitudes and reactions of the individual'.

The 1974 Act makes a contribution to this side of the problem by requiring that firms shall write down and publish their safety policy. The Act paves the way for greater consultation through safety committees and emphasizes the importance of information, instruction, training and supervision. It is not immediately obvious how a legal document could go further in tackling a problem which is only partly legal in nature.

So much for the background. There remains the problem of what to do on the factory floor, the construction site, and on the drawing board. Let us assume that all things required by law will be tackled in an honest attempt to keep within the law. We are not therefore concerned here with the problems of compliance with particular codes or regulations. These may be difficult enough but in general the problems are technological and too specific for detailed discussion in this volume. The project manager however will want an assurance that on his project, as far as possible, all the relevant codes and regulations have been observed. He will gain this assurance from a combination of his own personal experience, his reliance on the quality of the staff responsible for any part of the project and his introduction of appropriate independent checks. The balance between these methods will depend on the nature of the work in hand and the consequences of an error should it be allowed to pass unattended.

Over and above this legal requirement imposed by the Act on those who design or supply any article for use at work (Section 6(1)) there is also the problem of foreseeing the unexpected. In a sense this is also a legal requirement in that the legal requirement to make something safe does not stop at the compliance with regulation—rather the requirement is that the person involved shall ensure safety 'as far as is reasonably practicable'. It would be an interesting point to debate how far it is possible to foresee the unexpected!

In an attempt to discover hidden snares independent reviews of the project are invaluable. Recognising the adage that 'it is impossible to see the wood for the trees' such a survey at least increases the chances of protection against oversight. A particular technique of Design Safety Survey is illustrated in *Figures 8* and *9*. *Figure 8* shows why at certain stages in the design process, one should stand back and take stock of the situation. The three points chosen are significant because

The Anatomy of a Project

they immediately precede important developments in the creation and operation of the asset. Surveys 1 and 2 are undertaken immediately before significant increases in the design activity. Survey 3 is undertaken prior to production.

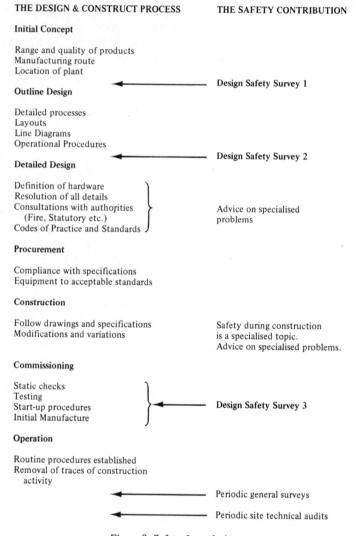

THE DESIGN & CONSTRUCT PROCESS   THE SAFETY CONTRIBUTION

Initial Concept

Range and quality of products
Manufacturing route
Location of plant

Outline Design   Design Safety Survey 1

Detailed processes
Layouts
Line Diagrams
Operational Procedures

Detailed Design   Design Safety Survey 2

Definition of hardware
Resolution of all details
Consultations with authorities   Advice on specialised
   (Fire, Statutory etc.)   problems
Codes of Practice and Standards

Procurement

Compliance with specifications
Equipment to acceptable standards

Construction

Follow drawings and specifications   Safety during construction
Modifications and variations   is a specialised topic.
                                        Advice on specialised problems.

Commissioning

Static checks
Testing
Start-up procedures   Design Safety Survey 3
Initial Manufacture

Operation

Routine procedures established
Removal of traces of construction
   activity

                                        Periodic general surveys

                                        Periodic site technical audits

Figure 8. Safety from design

Figure 9 is a more detailed description of the first of these surveys and has to be taken in conjunction with appropriate check lists.

Design Safety Survey 1

Timing
This should be undertaken when the concept of the scheme has been clarified in terms of the range, quantity and quality of the products to be manufactured, the outline process or route to be employed and the location of the plant has been chosen. Design work will not have begun except possibly for feasibility studies and no major commitment will have been incurred beyond the development work.

Purpose
The primary purposes at this stage are:

1. To verify that the concept of the scheme is right in terms of the selected environment.
2. To identify the areas of particular concern from the point of view of health and safety so that subsequent design work can be properly directed.
3. To provide a check that the proposals with regard to health and safety are acceptable or are capable of being made acceptable before further financial commitment is undertaken.

Scope
The examination will inquire into:

1. The materials used in the process.
2. The materials formed at intermediate stages.
3. The nature of the final range of products and by-products.
4. The nature and size of the equipment required.
5. The conditions under which the process takes place.
6. The degree of human involvement in the process.
7. The requirements of maintenance.
8. The layout proposed in terms of environment, scale and access.

Method
The precise method will vary according to the nature of the proposals but would include scrutiny of the available information with regard to materials, plant and location. Check lists will be used as aids to probing for unforeseen problems.

Report
The survey will include a report which comments on the findings and makes suggestions for consideration or action where necessary. The report would be confidential to the client.

Figure 9. Design safety survey

Safety and Statistics

In recent years there have been some bold attempts to quantify safety. These are welcome because they constitute a serious attempt to deal more effectively with this difficult problem. Recognising that there is no such thing as absolute safety, the problem in design often lies between two (or more) alternative courses. It is likely that the comparison can be quantified in terms of capital cost, operating cost, time for implementation and perhaps even with respect to incidence of failures. The best judgement that can be made on safety may well be subjective.

In these circumstances the scientific mind turns naturally to the problem of quantifying safety to see whether numbers can be assigned to measure the risks that are involved. There is unfortunately still a shortage of reliable data although the U.K. Atomic Energy Authority has devoted considerable effort to this problem. The chemical industry has also been the scene of much effort.

To quantify this matter a yardstick has to be adopted and one such is the fatal accident frequency rate (FAFR). It will be seen for example that one can compare two different methods of travel by comparing the appropriate fatality rate per ten million passenger *hours*.

Figures quoted are:

Travelling by bus	3
Travelling by train	5
Travelling by car	57
Travelling by air	240 etc.

These figures are helpful. They do not necessarily prevent us from flying or travelling by car but they give us an indication of the price we are paying for pleasure or convenience.

One must always remember that figures can be twisted more or less at will. If we assume that a car travels at 40 m.p.h. and an aeroplane at 350 m.p.h. then the two fatality rates measured per ten million passenger *miles* will be:

| Travelling by car | 1.42 |
| Travelling by air | 0.68 |

– a different picture altogether.

The problem of this approach is that one may sometimes have to choose the more hazardous solution. An economic motive is tantamount to ascribing a cash value to a human life, which is clearly distasteful.

The more tenable objection comes from the nature of the statistical approach. It may be possible to calculate that a certain type of explosion will happen once in a thousand years. This may be mathe-matically sound but it does not mean that it will not happen tomorrow—only that the risk is remote.

It must be remembered that human error plays a large part in many accidents. To what extent can the 'accident prone-ness' of an individual be measured? To what extent is an installation less safe because Joe Bloggs is working nights this week? Pity Joe Bloggs if anyone ever succeeded in putting a meaningful figure on that!

The nature of the project is that decisions have to be made in the light of all available information—sometimes quickly but never reck-lessly. If statistical information about accidents is going to mean better decisions then surely there is an unanswerable case in favour of making numerate assessments wherever practicable.

CHAPTER 3

Management Needs

At the beginning of this chapter we shall take a different stance and examine some of the features that management will require to be able to bring the project to a successful conclusion.

Objectives

We have referred to objectives in Chapter 1. At this stage we must stress the need for clarity because during the life of a project, especially in its formative stage, there will be many storms to blow the ship off course.

Determination to remain single-minded in the achievement of the objectives may have to be tempered by a willingness to stop or retrace steps if need arises. One must anticipate problems initiated by friend and foe alike and it is important that the objectives remain clear. We have already made the point that objectives should be written down. Perhaps we should now add that they should be engraved on the Project Manager's heart!

We should also see the objectives of a project as the basis for all types of management decisions. We shall be discussing organisation, responsibilities and resources in the context of project work. For every new project there are new problems to solve. Sometimes this means acceptance of the status quo but more frequently it means change. Change implies risk and therefore needs careful control. The sheet anchor from which problems are solved and decisions taken is the stated objective (or objectives) of the project. The basis for monitoring and controlling any changes which the project seems to demand will invariably be the expressed objective.

Organisation

To some the establishment of the correct organisation is the corner stone in tackling any task. This attitude denotes the tidy mind and in its way is to be applauded, but it is not by any means sufficient. One is reminded of the quotation from Robert Townsend's book *Up the Organisation* which should be read by every manager who might possibly be blinkered. It reads:

And God created the Organisation and gave It dominion over man
Genesis 1, 30A, Subparagraph VIII

The purpose of an organisation is to facilitate the tasks of people and when several are involved to make it possible for their activities to be coordinated. To expect the organisation to do more than allow people to perform jobs is to make an unwarranted assumption.

To use some simple and homely examples, if I undertake to cut the grass on my lawns I do not need (or wish!) to be organised. It is true that the task needs to be organised and planned to some extent because I need a mower, some petrol for the mower, maintenance, etc. But as far as an organisation is concerned none is necessary.

On the other hand if I join with three friends to run a car club to get us to the office each morning it begins to be obvious that we need one of us to make sure that the system is workable each day. We might appoint one of our number to be the organiser.

As more people get involved the need becomes greater and so one finds that larger bodies in social spheres have secretaries and treasurers and other nominated individuals to perform agreed functions.

In each of these cases the purpose of setting up an organisation is to permit people to undertake the work that they have to do in as efficient a manner as possible. The establishment of an organisation implies that the necessary resources are brought into a relationship with one another and that there is satisfactory communication or flow of information between them. It is also implied that the activities required are brought under satisfactory control to achieve whatever purpose is required.

Reverting to project work the position is the same. Organisation is necessary to permit the task to be undertaken efficiently. This is true whether the arrangements set up are part of a permanent system which covers many projects or whether the arrangements are just for the duration of a particular series of events.

It is true whether all the participants are paid from the same purse or whether some are brought in on a contractual basis and paid through a different mechanism (see Chapter 4).

Types of Organisation

(a) Military

The type of organisation shown. in *Figure 10* may be slightly unfair to the Army but it displays a clear-cut no-nonsense type of organisation. It is particularly suitable for a well-disciplined group required to perform clearly defined tasks in an efficient and orderly manner. It does not permit or encourage a great deal of independent activity.

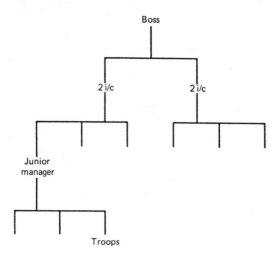

Figure 10. Military style organisation

In the field of capital projects it may well be the most suitable type of organisation for the smaller projects which are well defined and unlikely to encounter major deviations. This is especially true if there is an existing organisation (for whatever purpose) which can readily absorb the volume and complexity of the work which the new project requires.

(b) Matrix

Many publications in recent years have extolled the virtues of the Matrix type of organisation which is illustrated in *Figure 11*. Some illustrations are a great deal more complicated but in essence this type of chart recognises that a man may have a split responsibility. For the performance of certain duties he will be responsible to his line manager but for the *way* in which he performs he has a responsibility to someone else. It is conceivable that he might also have a third 'boss' for some aspect of his work but at this stage one begins to pity the man.

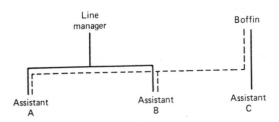

Figure 11. Matrix organisation

This arrangement has been developed to cope with the growth and increasing complexity of technology. The inference is that the line manager does not need to be an expert in the fields in which his assistants have expertise. Whilst he will be responsible for their work he will not necessarily be in a position to judge the technical content of that work and he will not necessarily be in the best position to know if it is technically up to date. In these circumstances a technical manager may judge the adequacy of the technology and be responsible for training, development and other technical functions.

This type of organisation is very suitable for a great deal of project work. As an example, a new process plant may incorporate a computer for on-line control of the process. This would demand specialised advanced technology. The member of the project team who was particularly responsible might need guidance from the technical manager on certain aspects but the time scale and the cost control of his work would have to be consistent with other parts of the project and for these aspects he would be responsible to the project manager.

If a matrix organisation is to be established it needs to be done with care because the people in the matrix will need to understand their position clearly. Such problems as salary, working conditions, job status and similar matters need to have a method for their solution or else the system suffers.

(c) Circular

A third way of expressing an organisational arrangement is shown in *Figure 12*. Some may throw up their hands in horror at this because it does not answer questions of responsibility and it does not adequately deal with the position of individuals.

Nevertheless it has a great deal of value in showing how central to the action is the role of project manager. The chart really assumes that there is in existence a form of organisation which is loosely 'military' in style but which is capable of being adapted, on a temporary basis at least, to serve the best interests of the project.

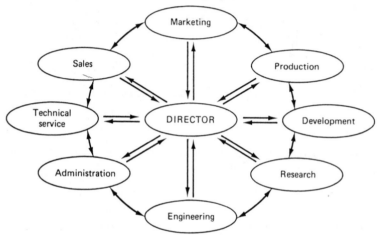

Figure 12. Circular organisation

Figure 12 is particularly appropriate because an early stage of project life is portrayed. The activity at this stage may well be open-ended in that full definition has not been achieved. A high degree of communication is therefore required. As the project develops with the roles of individuals becoming more clearly defined a more orthodox structure may be adopted.

The Operation of 'Teams'

Recent years have seen a great deal of experimentation into the question of tackling problems by team effort. We cannot move from the question without considering some aspects of this approach.

There is nothing new in the idea of tackling a problem by the application of team effort. The twelve tribes of Israel were independent units until they needed to join together to fight a common enemy or in the early days flee from Egypt under the leadership of Moses. By joining together they achieved a purpose which would have been impossible as individuals.

What have changed in recent years are the technical complexities of the jobs to be tackled and the time scale in which they are to be performed. This means that it is very difficult for the 'boss' to be the superior of all his men in all respects. This, as already indicated, is the reason why 'matrix' types of organisation have tended to replace the 'military' types in project work.

But this alone is not enough because in project work all the efforts of the individuals engaged on the task have to be focused onto the project objectives. The project team is the means of focusing these efforts. The size of the team is not necessarily fixed but it should not be so big as to be unmanageable. The members of the team do not need to be of equal seniority or status although each must recognise that others have something to contribute which they personally cannot give. This gives the team members some mutual respect which overcomes problems of different backgrounds.

The purpose of a team will be to solve a particular problem or achieve a stated objective. This might be for example to undertake a feasibility study or a development programme. Reference to *Figure 16* shows that interdisciplinary activity is usually highest at the evaluation and early design stages and these will be the stages where teams are most appropriate. Once the purpose of a team has been achieved it should be disbanded because its members will be required for other work for which they need other contacts.

The team does not have to operate on a full-time basis but it should essentially be an arrangement at the working level. It is not satisfactory to form a team of heads of department who meet once a month and delegate the action to members of their staff. The team members in this case should be the appropriate members of staff. They will benefit much more from rubbing shoulders with people of other disciplines than if this is done secondhand through their bosses.

Leadership of teams has been tried in a number of different ways—by a fully active member of the team, by an extra member of the team brought in to be leader, or even by a more senior person outside the team leading on a part time basis. The important consideration is that the team should have clear objectives and a plan to achieve those objectives. The leadership function will be to ensure that the objectives and plan are established, remain valid even in the light of the work of the team, and are finally achieved. Again, this requirement will be true whether the team is tackling a large problem or a little problem and whether the duration is a month or a year or more.

Leadership will be seen to depend largely on the personalities of the people involved. In this case there is no golden rule to say that one particular method of leadership appointment should always be employed. The ability to command respect, the experience available and questions of temperament and responsibility will all contribute to the choice to be made.

In certain circumstances the Project Manager himself might very well be chosen as the leader of the team. There may however be several teams contributing to the project activity and sheer volume of work may demand that the project manager is not available for this particular task. The Project Manager however must exercise a coordination role even if the main task of the team is delegated to others better able to perform satisfactorily.

It is perhaps necessary to draw a distinction between a team and a committee. Although they might be similar in that they both have objectives to fulfil, their method of working is likely to be entirely different. A typical committee will have rules and procedures, standing orders to abide by, elections and minutes. None of this formality is necessary in a team activity and should not be invented unless there is a very good and *real* reason. It will normally be sufficient to have notes of decisions taken where appropriate just to avoid misunderstandings and action lists if it proves necessary to help team members with their individual planning.

Team decisions should preferably be unanimous and corporate. Individuals should not be afraid to stand up for and by the corporate decision if necessary. Equally a team member should not be judged by his superior for a team decision which that superior does not favour. It must be recognised that when a member of a team contributes to a decision he does so to the limit of his ability but he may be persuaded by considerations advanced by the other team members who will be contributing to the same decision making process. It is most damaging to team operation to have one or more of the team members influenced by what his boss would do in similar circumstances or running to ask his boss before he says anything useful.

Responsibilities

It will be a prime requirement for the management of the project that all participants are aware of their responsibilities. This is true whether teams are established for specific tasks or not. No person can contribute to the maximum of his capability if he is not sure of the boundaries of the task which he is expected to perform. If there are omissions or overlaps between adjacent areas there will be neglect or duplication. Both impair efficiency and lead to frustration and lack of confidence. Both can be avoided to some extent by communication but time devoted to communication merely to solve responsibility problems is used inefficiently. It is better to foresee the difficulty and map out the areas of responsibility in the first place.

Management of the project demands that responsibilities are clarified also to make sure that action will take place on all the required aspects. Approval of a project for example might depend on a satisfactory appraisal of possible alternatives to demonstrate that they are not preferred. Such work might appear to be negative and unpopular to those who undertake the task. Nevertheless it is important, and it is vital that the results are available at the right time.

Alternatively for example the project may involve a complex legal issue on which professional advice is being sought. If the Project Manager arranges for outside help to be obtained he would not wish some other part of the organisation to stumble over the same problem and attempt a separate solution. Clarification of the responsibilities would help in such a case.

Communication

Special attention needs to be paid to the problems of communication. Many excellent people are indifferent communicators. Too little communication breeds assumptions (which must sometimes be wrong) and inefficiency on the part of the receiver. Too much communication implies that some of it is irrelevant and the receiver switches off at the wrong moment or fails to attach the right significance to some part of the message.

The golden rule is to remember that all communication is for the benefit of the receiver. He needs it so that he can respond correctly to what he receives. If he ought not or does not wish to respond to any communication which he receives, he would be better without it for comprehension only absorbs unnecessary time and energy.

Communication of course can be written or verbal or some combination of the two. If written there are a variety of guises under which the same thing can be said. We use letters, memoranda, reports, minutes of meetings, action lists, drawings and many other vehicles for disseminating ideas. It is important to choose the best system and the best will be that which uses the fewest words to give the right message.

Copies should be used with discretion for they take as much time to read as originals and therefore can sometimes become disregarded.

Remember also that communications are not always easily understood. The receiver has a choice as to whether he reacts actively or passively to the message which he receives. Was the message for action or for the information? It is useful to put oneself in the shoes of the receiver and try to anticipate his reaction.

Information

Lest the previous section should be too narrowly interpreted as meaning just simple messages we should look at the type of information which is likely to be used in project work.

Basically this of two types:

(a) that which relates to the viability or the assessment of the project, i.e. the external or global view of what is going on.

(b) that which is necessary for the execution of the project, i.e. the internal workings of the system.

Of the first type, we discussed the information which would be necessary in a public sector project in the swimming bath example in Chapter 2. A great deal of work has been done on this subject in the private sector and Appendix 2 on Risk Analysis contains references to many items on which information is required.

Execution of the project will require information which in turn can be subdivided to indicate the range of possibilities.

(a) Technical information expressed in drawings, specifications standards, instructions or reports.

(b) Commercial information such as enquiries, orders, conditions of sale or contract, measurements, prices etc.

(c) Administrative information such as letters, memoranda, permits, agreements etc.

Information of both types may have different significance with respect to time. Some will be ephemeral and some will be of lasting importance. This time aspect needs to be recognised when it comes to questions of filing, information retrieval and its general availability. If Jack has a cold and won't be in until Monday that information has already diminished in importance by Tuesday. In the long term it may not be significant at all or only to the extent that so many man hours were lost. Deciding what information should be kept and how and where may sound like an administrative chore but on a big project it can become very important. The savings on filing cabinets and clerical time may be significant but most important of all could be that the really useful information becomes lost in a mass of trivia.

Approval is another question which needs attention. Drawings, calculations, orders, etc, should receive appropriate authority before commitments are made. This may well be universally recognised in theory but is not always observed in practice. 'Rubber stamps' are sometimes made with a fountain pen as well as an ink pad!

We shall be saying more about control but it is here merely necessary to note that as projects grow in size, and personal contacts are diluted, the information systems employed increase in importance. This is not to advocate excessive formality but to point out the danger of the system taking over. There really is a lot of advantage in having a system which is 'appropriate' to the project rather than a legacy from the last one.

Project Data

In one sense project data is the link between the two types of information which are found in project work. By definition Project Data is the continuously developing basis upon which the project is brought to completion. It is a step-wise clarification of the project objectives.

The project moves from its initial concept and statement of objectives through stages of development and optimisation leading to detailed design and construction. Through this movement there is a transfer of information from that which is used for assessment and appraisal to that which is used for actual execution of the project. Thus the information which forms Project Data is a very special part of the second type noted above.

The concept of project data is an important one and Appendices 3 and 4 are devoted to examples of project data in the private and public sectors respectively. The concept is illustrated in *Figures 13* and *14*. *Figure 13* shows how, during the early stages of a project, a broad statement of the project requirements is brought to focus on specific proposals. This work is simultaneous with the build up of data which is needed for subsequent design.

To choose a commonplace example of this clarification process let us consider what happens when we purchase a car. The idea that we might purchase is the beginning of the story. We shall next seek justification in terms of economic viability (can we afford a car? how

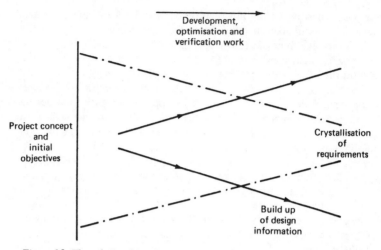

Figure 13. The relationship of design information and clarification of need

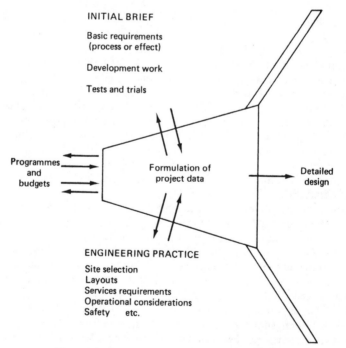

Figure 14. The relationship between data and design

much will it cost to run? how much will it save us on business or on holiday? etc.) We may also review other benefits (nice to have, comfort, keeping up with the Joneses!). Gradually as we review the various factors there comes an emerging picture of the car (how big? what make? what colour? when?) The requirements are crystallised at the same time as we have built up a picture of what we are going to do, i.e. buy a particular car from a particular dealer at a known price. Also we shall have considered problems of finance, garaging, maintenance and perhaps what to do with the old car.

This example of course covers very nearly the whole action of this particular project but in major capital schemes there is a very substantial design effort for which preparation must be made. *Figure 14* shows this as a funnel through which all the project requirements in terms of objectives, standards, programmes, etc, are fed into a preparatory phase which eventually permits the project to move forward economically and quickly.

Double arrows are shown in this illustration to indicate that some backtracking is to be expected. In fact in complex, high-risk projects the amount of backtracking will be considerable. This is not necessarily objectionable at the Project Data stage because some trial and error would suggest that improvements are being sought. It is however important to keep the backtracking (which is like playing snakes and ladders) out of the detailed design stage as far as possible because at that stage many more people are involved and difficulties of communication would suggest that risk of error is greater. Detailed design of steelwork, concrete, pipework or electrics will bring its own problems and is preferably free from fundamental changes which radically affect all that is being done.

The Development Stages

Project Data will be prepared during at least part of the development stages. It is important that these early stages of project work should be brought into the total scene. It is worth looking in more detail at this area, which entails the action from the concept of the idea through to the point at which it is accepted that a particular project will proceed and be built.

It is a characteristic of this area that there is a high mortality rate. Many promising ideas are dropped on closer examination either because they are not attractive commercially or because the technical problems are too great (or too risky).

There is no way of avoiding a high mortality rate. Indeed paradoxically it is a sign of health because every organisation needs a plentiful supply of ideas if it is to grow and expand.

The hope is to pick the winner at an early stage so that it can be nurtured and accelerated to maximum advantage at the earliest possible time. The problem will be to evaluate its potential as early as possible so as to increase the certainty of knowledge that it is a winner. The evaluation process itself costs money and therefore needs, if possible, to be executed simply and cheaply. As a first test, 'hunch' or 'experience' is usually applied. There is nothing wrong with this, but observe that many managers are happy to reject ideas on this basis but not so happy about knowing what to do with the ideas that they cannot immediately reject.

We are now going to suggest that if the idea is not rejected on a first appreciation and if it is seen that capital would be involved then that idea should be immediately labelled as a possible project. If this is done then the idea will not be lost or forgotten. Nor will it remain in the 'Too difficult' box. Instead it will come up for formal review and its future will be a matter of deliberate and conscious choice rather than neglect or fortuitous support. Only in this way will the costs of evaluating that idea be kept within bounds and, if the idea does prove to be a winner, it will be pushed as quickly as possible through the evaluation stages.

The evaluation stages of course will follow the general pattern illustrated in *Figure 3* and described in Chapter 2. At this point it is only necessary for us to consider the work content of those evaluation stages. This may take the form of market research, desk or paper studies, laboratory work, pilot plant scale work or perhaps a development programme on the full scale proposal. The particular work required will be a function of the problem.

An example in which a series of experiments are designed to test an idea is illustrated in *Figure 15*. This illustrates the logic which goes into a series of actions. If the experiments have significant time scales, (e.g. six months), it will be seen that the success or otherwise of the programme will have a marked effect on the time scale of the project. If A and B fail but C is successful the programme is already 12 months behind the earliest possible, and still faces a new test programme which itself will take a finite time. Careful planning will not alter the technical outcome of A, B or C but it will help the decision as to whether to embark on the work or abandon it much earlier.

The common objection to planning in the development stages is that it is not possible to predict how long it will take to think up a new idea. Nevertheless the attempt to set up a logical sequence of events is of itself valuable and may indicate immediately that it is not worth trying to think up a new idea because any possible outcome would not fit into the total pattern.

It is not only time which is important but also resources. In the public sector in particular time may not be a critical feature in the assessment of a particular benefit. The use of resources is more likely to be important either because the resources required are simply not available or because they are expensive to employ.

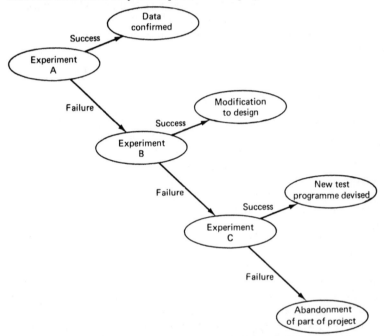

Figure 15. Planning logic for development work

There is a real difficulty in tackling this problem realistically when the organisation concerned (be it public or private) uses in-house resources for development and evaluation work. Because in-house resources tend to be considered as (comparatively) permanent they are costed as a fixed annual charge. This charge is then incurred irrespective of the results. Under this system it is very difficult to allocate costs effectively to particular projects and even more difficult to take decisions as to whether a given piece of work is justified or not.

This is not intended to denigrate in-house organisations for they have a function to perform, particularly where highly specialised work has to be done. It does, however, underline the importance of decision making in early project stages. It may sometimes be convenient to do a parcel of work because resources are available, but the true cost of such resources and the price of the effort expended should be recognised and identified against the criteria by which the project as a whole is being judged.

Otherwise, as far as the development stages are concerned, much additional attention must be paid to the matters discussed in Chapters 4 and 5. The development phase is just as much a part of the project as subsequent detailed design or construction. The rate of expenditure may not be so high but the effect on the overall outcome of the project makes the control of this phase most significant.

Forecasting

We have discussed forecasting to a limited extent in the Overall Financial Assessment of Chapter 2 and of course the Appendix on Risk Analysis is relevant. We mention this important topic here to draw attention to the need of management to understand the problems of forecasting and that there is only a probability of any forecast being right. There is no way of predicting the future with complete certainty and, in the commercial world especially, the chances might include some long odds.

People themselves are unpredictable so any result which depends on personal choice is immediately open to some doubt. This is not a cynical view. Political events, for example, often impinge in some way on major capital projects and there are the customary uncertainties of economic trends, public taste, commercial pressures etc.

All the uncertainties combine to make forecasting a difficult subject and yet it is of considerable importance. All decision making is intended to affect future events and decisions are normally choices between several courses of action. The decision will be based upon the forecasted consequences of each particular course of action and thus the correctness of every decision will depend on the accuracy of the forecasts.

Historical evidence is only partly helpful in making forecasts because, after every decision, the events which result cannot really be compared with the events which might have resulted.

Some managers therefore make decisions by hunch and intuition on the basis that making the right decision is more important than using the right techniques of decision making!

Be that as it may, complex projects demand a more rational approach because the financial implications of error are so severe. Maximum care is demanded and this is usually achieved by applying some of the advanced mathematical techniques which are available.

CHAPTER 4

Planning the Action

Resources and Contracts

The achievement of project objectives will normally require the application of considerable resources. As a broad simplification all expenditure is for human effort or material. We recognise also that most engineering or building materials, e.g. steel, aggregates, etc, are comparatively cheap. It therefore follows that a high percentage of the total money involved will be spent on the provision of labour. This can be split into three categories although the proportions in each category will vary considerably between projects. The categories are as follows.

(a) *Design and Investigation*—This group will include not only all the draughtsmen and engineers needed to design the project but also all earlier stages. Market research, development programmes, feasibility studies, negotiations and financial assessments may all be required and can demand considerable effort with a high degree of interdependence which may cause confusion and complication if not properly coordinated.

It is also true to say that the labour required at the early stages of a project will demand a wider variety of disciplines than at the construction stage. Not only will there be the common engineering disciplines of Civil, Mechanical, Electrical, Instrument, etc, but also the Accountant, the Lawyer, the Planner, the Market Researcher, the Academic Scientists, etc. *Figure 16* gives an indication of the way in which demands vary during the life of a project.

(b) *Site labour*—This may well be the largest category in numerical terms and it is characterised by the brevity of duration of any one type of action. It will include all the specialists, tradesmen and labourers who are required to assemble the job on site.

(c) *Off-site labour*—A lot of the equipment, plant and probably some of the building materials will have absorbed labour before the items

51

concerned reach the site. As an example an item such as a fork-lift truck will be delivered to site ready for use except perhaps for a post-delivery check and service. On the other hand items such as bricks have a comparatively low off-site labour content.

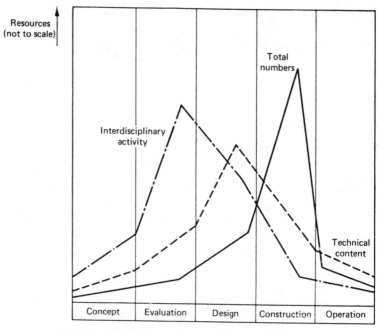

Figure 16. Inter-disciplinary activity in the life of a project

Leaving aside off-site labour for items which can be described as standard stock items, all functions of these three categories must be realised in the time between the clarification and approval of the project objectives and their achievement. Usually this interval is short and often there is an alleged advantage in making it even shorter. The basic problem is to mobilise these resources effectively to contribute to the achievement of the objectives. In setting about this several factors have to be considered.

Motivation—This might not be the first item in everyone's list, but it ought to be. The project needs draughtsmen and craftsmen, engineers, managers, tea boys and testers. If the project is to be really successful their motivation has to be brought in line with the objectives of the project. They cannot be expected to have identical objectives because in the nature of things their associations with the project will vary as to time and content. Nevertheless the objectives must harmonise and the individual motivation must be compatible with the achievement of the project objectives.

Experience—Experience usually determines how a particular job is done. Choosing the right method usually allows a job to be done efficiently (which means cheaply), quickly and safely. Thus the right choice of method will almost certainly contribute to the achievement of the project objectives.

In case this should be thought of as an oversimplification let us acknowledge that in this aspect there is room for judgement. Skill is expensive and may be scarce. It may therefore be expedient to use a less skilful resource on a particular task. The problem depends on the individual circumstances but for any one task there must be a minimum level of experience which is acceptable.

Availability—Having recognised the need for certain resources at a certain time, are those resources truly available? This is sometimes a very sore point when a particular specialist skill is in short supply—to such an extent that it is not uncommon for deception to be practised. It has been known for mythical lists of people to be conjured up in order to satisfy a casual enquiry.

The real problem faced by those who employ labour of any discipline is that they require enough work to keep that labour fully occupied— no more but certainly no less. Since the search for work has varying degrees of success, it is not always predictable as to how this will affect the future and the duration of the job under consideration.

This labour availability question deserves sympathetic consideration on the part of the employer. Flexibility should be built into the plan but it will need continuous review because availability can and does vary significantly during the life of a project. There is nevertheless a minimum level of resource requirement which must be established at the outset and the true availability must be measured against it.

The point is perhaps best illustrated by an example (*Figure 17*):

Let us suppose that a task has to be performed which requires 160 man weeks of tradesman effort. The original programme for this task requires it to be performed in 10 weeks but a close examination of the programme shows that it is on the critical path if it takes 16 weeks or more. Several things can then be stated about the resource availability which is being sought:

(a) The very minimum number of men required on the job continuously is 10. At this level there is no margin for variation or error and no room for gradual build up or run down.

(b) The safe minimum is 16. Not only is this the theoretical number required if the job is really to be done in ten weeks but also it represents approximately 50% over the theoretical minimum required

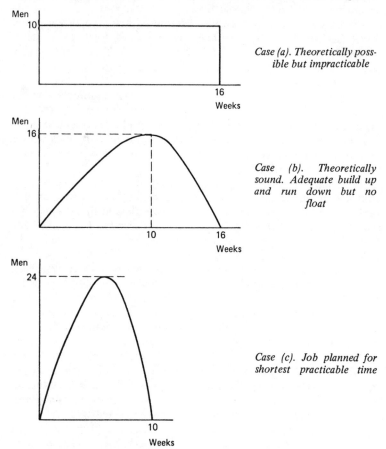

Case (a). Theoretically possible but impracticable

Case (b). Theoretically sound. Adequate build up and run down but no float

Case (c). Job planned for shortest practicable time

Figure 17. Different ways of meeting requirements

to do the job in 16 weeks. This 50% excess will allow a margin for winding up and running down and is likely to take care of casual turnover during the job.

(c) The preferred available resource is 24 which allows for the 50% excess referred to above and still expecting the job to be done in 10 weeks.

In practice a real availability of less than about 13 would be rejected as inadequate for the task. If the number were between 13 and about 20 further safeguards would be sought. Only if the number were above about 20 would reasonable confidence be engendered. In all cases however the real availability should be continuously monitored during the life of the job.

The availability of equipment and facilities must also be considered. If the equipment required is specialised it is unlikely that additional resources can easily (or cheaply) be mobilised at a late stage in the action. Very heavy cranes for instance tend to become earmarked for specific jobs and short notice requirements may be difficult.

Management structure—Following on from this subject in the previous chapter there is no particular structure which is right in the sense that all other structures are wrong. The question to be asked is whether the management structure of the team of resources in view will be adequate (*a*) to mobilise those resources effectively at the right time and place (*b*) to control those resources to contribute to the achievement of the job in hand and (*c*) to liaise adequately with other organisations involved in the action.

To some extent the answer to the question is to be found in the experience and track record of the management team. If the team has achieved a lot of similar work in the recent past there is a clear indication of the probable success of the present task. It must however be remembered that no two projects are the same and the situation to be faced on the proposed task may vary in some crucial respects from previous performances.

In particular it should be remembered that legislation, government policies, economic factors, union pressures (and militancy) are all liable to change. If the project in view is of considerable size it will be all the more important that the management team are aware of and able to cope with variations in external circumstances.

Financial security—Firms whose business is to provide resources for capital projects have normally established methods of ensuring financial security by way of bonds and guarantees.

The user of the resources has one main interest in this field. He does not wish any supplier of resources to fail to meet his commitments. On the financial side he will therefore be concerned to establish, as far as can reasonably be predicted, that shortage of ready cash is not going to prevent resources being mobilised for the completion of the project. He will wish to make sure that there is a wide margin of safety against this happening because in a sizeable contract many events may contribute to a squeezing of the cash. To name a few possibilities, these might include:

> Adverse weather causing delays.
> Accidents causing damage to person or property.
> Delivery of material problems.
> Subcontractor problems.
> Unsatisfactory workmanship.
> Labour disputes, etc.

Matching Resources and Needs

A very broad simple split of the work to be achieved is shown in *Figure 18*, where possible resources are also listed. It must be remembered that the dividing lines between the work headings are not sharply defined and certainly overlap in time. Thus preliminary studies soon overlap with and become design. Also some design, especially detailed drawings, may well be done by the construction contractor even though nominally the design may be executed by others.

(a) The client of course may elect to do the complete job with his own resources. An organisation which is geared to a fairly steady programme of capital development will be able to do this provided no special problems of expertise are involved and provided the time scale of the work is such that the available resources are not overloaded.

If the client has such an organisation there will be many debates about its size in relation to the future work load. These debates will include comments on the cost of 'in-house' design versus use of outside resources. One suspects that there is no answer to this conundrum but it is a sharpening exercise to keep the subject under debate.

We need only note here that if the client's organisation exists and has available capacity and competence it would be uneconomic in the extreme not to use it.

(b) The use of consultants includes a wide range of professional talent which is available. Although the major activity of the consultant is thought of in terms of design the role can readily be extended back into the conceptual and formative stages and forwards into the fields of construction supervision and quality assurance.

Consultants, while not undertaking construction activity, are able to give the client impartial advice on this subject. Nevertheless they would normally take the necessary steps to establish a contract between the client and the construction contractor so that the job can proceed.

| Possible resources | Preliminary studies | The Work to be achieved | | Construction |
| | | Design | | |
		Basic	Detailed	
a. CLIENT (OWNER)	Necessary	Possible	Possible	Possible
b. CONSULTANT	Assistance possible	Major activity	Major activity	No
c. DESIGN CONTRACTOR	Assistance possible	Possible	Major activity	Possible
d. DESIGN and CONSTRUCT PACKAGE	Assistance possible	Possible	Major activity ←— Link	Major activity
e. MANAGEMENT CONTRACTOR	Assistance possible	Possible	Major activity	Major activity including subcontracts
f. CONSTRUCTION CONTRACTOR	No	No	Possible	Major activity

Figure 18. Different contractual arrangements

(c) Design contractors, large and small, have emerged as a result of larger and more complex industrial developments. Many are linked very closely with construction organisations, and this is possibly the reason for their being there in the first place. They can thus hardly be said to be impartial although most are willing to undertake design tasks without necessarily entering into any commitment on the construction aspects.

The larger design contractors usually have a number of professional chartered engineers on their staff and as a general rule a good quality performance is achieved. Their strong point however is the commercial production of design information for large jobs rather than the solving of difficult or novel problems at the design stage.

Most design contractors operate on a time-reimbursible or fee basis rather than on a percentage of the completed work.

(d) There are an increasing number of firms who will offer a design-and-construct package and from the client's point of view there are enough advantages to make this method worthy of closer examination.

(i)   The communication problem between design and construction teams is minimised since both teams are employed under one contract. This eliminates a source of claims from contractors who do not have design information or who have to do work twice because of wrong design information.

(ii) The method has the *potential* of faster overall completion.

(iii) A package deal of this type may be easier to establish and administer since there is a common management for so much of the action.

On the other hand there may be disadvantages:

(i)   The combined design-and-construct resources may not be so experienced or technically well equipped as those available separately.

(ii) The nature of the contract makes it somewhat more difficult for the client to control, especially in respect of quality.

(iii) In effect the construction contract is being let before the design is completed and this limits the ways in which payment can be made for construction.

(iv) Many of the risks of the project (see Appendix 2) are still in reality carried by the client.

The design and construct principle is behind the arrangement in which a contractor is licensed for a process and holds the know-how required by that process within his own organisation. In this case the client expresses his requirements in terms of effects produced and agrees the consideration with the contractor. The satisfaction will normally be expressed in terms of a performance guarantee.

(e) The management contractor concept is best described as a special form of design and construct package. It is particularly applicable for very large jobs where the provision of adequate resources would otherwise be a problem. It thus becomes a rather specialised subject.

(f) The orthodox construction contract may be based on design work undertaken by the client, by a consultant or by a design contractor.

Paying for work done

The method by which payment is to be made must be determined in principle before a contractor is finally chosen. There are many variations possible but we will consider the most common methods of payment together with their advantages and disadvantages. The methods apply whether the task to be undertaken is design, construction or any other element which may be required.

(a) *Lump Sum*　This is the simplest form and is appropriate where the task to be performed is well defined, the problems likely to be encountered are minimal and the client's participation in the action is at a fairly low level. It has many advantages from the client's and the contractor's point of view because:

(i)　It allows both parties to understand the extent of the commitment before it is begun.
(ii)　It places the incentive for 'getting on with the job' in the contractors' court.
(iii)　It encourages the contractor to adopt a flexible approach to operate as efficiently as he can.

This is not a recommended method however unless the job is well defined by drawings (if the client is providing the information) or by agreed specification (if the contractor is offering to produce an effect). By definition there is no implicit agreement for the pricing of extra work and so it is desirable in these circumstances that extra work is kept to an absolute minimum.

A developer who cannot reasonably predict that variations will be minimal should not be advised to adopt a lump sum form of contract.

(b) *Schedule of Rates*–(i) *Construction*–The principle is that the cost of a given operation is agreed without necessarily knowing how many times that particular operation is required. Clearly approximate quantities are required or else the total size of the job in hand cannot be predicted with any accuracy. Also it must be recognised that there might be dramatic differences in costs between doing an operation once or doing it 100 times. The cost of piling a foundation is a good example of this. It is comparatively expensive to bring a piling rig onto a site. It makes a difference to the cost of the unit operation as to whether that rig is used for one pile or for a hundred piles. On the other hand the unit cost of painting (by hand) is unlikely to be significantly different whether the task is to paint $10\,m^2$ or $1000\,m^2$. It is relevant to note that recent work in the civil engineering field has introduced a distinction between method related quantities and time related quantities (see 'Further Reading').

The common practice is to prepare approximate bills of quantities. These would be used to establish the appropriate rates, the amount being determined by actual measurement of the quantities either on the completed job or from 'as built' drawings. In doing this it is recognised that if the quantities prove to be significantly different from the original prediction, there may be a case for adjusting the rates to be used. This would normally have to be done by negotiation–a practice which is much more acceptable in the private industrial sector than it is in the public sector.

This method of payment permits a pricing structure to be agreed between contractor and client before a lot of the detailed design has been completed. It thus allows a comparatively early start to be made on the job. Since it is inherently fair to both client and contractor it is a widely used system. It is capable of being used for a whole job or only for part of a job in conjunction with either lump sums or negotiation based upon cost plus.

It is appropriate to sound a word of warning on the use of this system as a basis of comparison between tendering contractors. It is unwise to examine only the total sum derived when the approximate bill of quantities is extended by the contractor's rates. The individual rates should be examined also. If a contractor, whilst tendering, has cause to believe that a particular item in the bill is grossly undermeasured he will tend to overprice that item even at the expense of underpricing on items which he believes are overmeasured.

If a contractor does this either by accident or design an unsatisfactory situation will follow. If the hunch was right the item(s) undermeasured and overpriced will cause the total bill payable to be greater than expected. The client, had he foreseen such a situation, might have chosen a different contractor or at least entered into discussion with the contractor to substitute a more accurate rate. If unable to do this the client may well feel aggrieved. If the hunch was not right the contractor will not be making money where he expected to and this also can have a damaging effect on his performance on site. Again this rebounds to the client's disadvantage in terms of the overall efficiency of the project achievement. The subject of contractor selection is of course deeper than this particular point and is dealt with on page 63.

(ii) *Design and earlier stages*–The same principle of rates can also apply to situations where the service purchased is essentially that of a man tackling a problem. In these cases the measure is made not by the results achieved because this is much less tangible than the hardware of construction. Thus it is common to pay for design by the hour or by the day, having specified the type of person who would be engaged on the task, e.g., engineer, draughtsman, tracer, etc.

Earlier investigation stages such as feasibility studies or market research can also be covered in a similar manner. In these cases the type of person required to undertake the work will also need definition and will probably require an extension of the above categories.

It should be mentioned that many professional bodies including members of the Association of Consulting Engineers, architects, quantity surveyors, etc. assess their fees on the basis of a percentage of the value of the work commissioned. This is a system which has worked well over many years and is particularly appropriate where the scope of work undertaken is traditional.

(c) *Cost plus*–The principle is that the contractor is reimbursed for all costs incurred together with an appropriate sum for overheads and profits. The payment for costs incurred is the same principle as that which applies if the job is undertaken by the employer's own staff (i.e. direct labour or 'in-house' work), thus there is no justification for the out-of-hand rejection of this method by ill-considered prejudice. There are however a number of safeguards which need to be considered.

(i) A good audit system is required. To avoid its being bent by the unscrupulous a satisfactory system of measuring for payment needs to be agreed between the parties.

(ii) The client should recognise that he has taken the profit motive away from an organisation which is primarily built around such a motive. This may well have wide repercussions at all levels of the organisation and the vacuum created will need to be filled by good management and possibly a more-than-usual interference in the contractor's organisation. To avoid undue problems the means of tackling the situation must be decided before a contract is agreed.

(iii) Over and above the costs which can be recognised and agreed the contractor will incur less tangible costs and would naturally expect profit from his operation. The less tangible costs may broadly be described as overheads and include such items as head office charges, insurance premiums, costs of idle services, advertising and publicity, etc. It is usual to roll all these up into a management fee which also includes a profit element. This fee can be framed in a number of ways, e.g.

> Fixed amount for the job.
> A percentage of the turnover.
> A sliding scale based on turnover.
> A variable fee (target fee) with an incentive element built in.
> etc.

If the contract is a large one and there is a temptation to work out a complicated method of calculating the fee. In general however as long as the method is fair and equitable there is a lot to be said for keeping it simple. It should be recognised that cost-plus contracts remove a great deal of the risk from the contractor's shoulders and he will therefore accept a lower profit target than on other jobs. Nevertheless he will expect and deserve a profit as on any other contract.

Cost plus contracts are particularly appropriate in certain special circumstances and these include:

(i) Emergency jobs in which special skills are called up without delay.
(ii) Jobs in which interference factors are likely to be so dominant that no other method of reimbursement is satisfactory.
(iii) Jobs in which the scope of the work required cannot be determined at the outset.
(iv) Highly specialised jobs demanding rare skills or equipment.

(d) *Variations*—It is foolish to pretend that any contract will not have its variations. To eliminate variations (which are seen as 'loopholes' by some clients) may be a praiseworthy target but it is never attainable. The only way to get near to this target would be to have the job so well defined and so well sewn up that unjustified delays would be incurred and the task to be performed would be insufferably late.

It is better to recognise that most contracts of any size are let with some loose ends and that some changes of intention are likely to arise. The target then is to make sure that there is a basis and a mechanism for agreeing variations as they arise. It is worth striving to get variations priced as they are originated because this eliminates a dangerous source of ill feeling.

Choosing the Contractor

For the sake of simplicity this section deals with construction only, The impact of these considerations on design and other activities will be considered in the next section.

The criteria for choice were dealt with earlier in this chapter and included:

> Motivation
> Experience
> Availability
> Management structure
> Financial security

Price was deliberately omitted because in general it should be very much a secondary reason for choosing a contractor. Indeed, contrary to common belief it is not really possible to judge which contractor will have done the job for the lowest price.

Consider the situation at the tender stage. Whatever the form of pricing employed the one thing that is certain is that the figure at the bottom of the last page will not be the amount of money actually paid at the end of the job. The tender in fact is only an approximate measure of the value of the job and its main function is to provide a satisfactory basis for payment during and at the end of the job.

If three (or more) firms have been invited to tender for a particular contract it is to be expected that the prices submitted will be recognisably within range of each other. If they are not it is worth enquiry as to the reason. This may reflect such factors as lack of experience or non-availability of the contractor's resources. It may also reflect misunderstanding of the contract documents or the degree of risk involved, or an inaccurate appreciation of the scope of work. If the former group of factors apply it was clearly a mistake to invite that firm to tender in the first place. If the latter group apply it will be worth considering whether the tender documents have in fact formed a satisfactory basis for tendering and whether the priced tender is going to be an acceptable basis for a contract.

It must be remembered that tendering is a costly operation for which a contractor gets no reward except on the contracts which he wins. If his success rate is one in ten then he has to recoup the cost of nine tenders as an overhead on the one contract which he wins. For this reason open tendering is not a system which can be commended. Since the invitation to tender is issued by or on behalf of the client the abandonment of the practice must be in his hands.

It is much better from the point of view of all parties concerned to use a selected tender list of three to six firms for each job. Inclusion on the list is obtained by pre-qualification in terms of the general criteria listed above at the head of this section. Thus if any of these criteria were unsatisfactory the firm concerned is precluded from tendering for this or future work until the shortcoming is amended.

It is most important that such a list is kept up to date if it is to be used for a succession of similar work. Thus it will not exclude new entrants into the field. It is also important that firms who decline in acceptability should be removed from the list. In compiling the list the client is really saying that he will accept any of the listed firms to do jobs which may arise. The final choice might still depend on factors other than price because the listed firms will clearly not be equal in all respects. Separate lists may well be kept for different sizes or types of job.

Alternately a contractor may be nominated for a particular task. This is a method which is rarely used in the public sector but which has an important place in the private sector. Quite simply the client or his representative equipped with a background of knowledge of the needs of the task and the way in which a particular firm is likely to perform makes a direct approach to the firm concerned with a view to arranging a contract. The selection procedure may be from a limited number of firms who may be interviewed to check their availability or suitability but in the end the choice of a particular contractor is made without competitive tender.

The basis for a price in these circumstances will be a build up of information from previous competitive tendering modified as necessary to suit the new circumstances. Clearly an atmosphere of trust is necessary for this method to be applied and this is a most important aspect of a negotiated contract.

The circumstances in which this method is likely to be employed include:

(a) where the nominated firm is a licensee for a particular piece of equipment or method required.

(b) where the availability of expertise or speed of operation is the main criterion by which the choice is being made.

(c) where a long-term relationship between two firms exists or is required beyond the terms of the contract under consideration.

The terms under which payment is to be made need to be determined in the same way as any other method of choosing a contractor.

Choice of Design, Development or Investigation Resources

In the earlier stages of the project the availability of adequate expertise is probably the most important single criterion affecting the choice of resources. It is more likely therefore that a nomination procedure will be preferred to a tendering procedure. This is particularly true as geographical and communication problems tend to be more damaging in the pre-construction stages.

If the price to be paid for work done under this heading is to be on a time-reimbursable basis it will obviously be necessary to determine what rates will apply. The warning to the unwary is that rates do vary quite widely. Also it is necessary to be sure of the meaning of terms. For example the word 'engineer' may mean a chartered engineer with twenty years relevant experience or it may mean a young man who has recently been given the title without any qualifications.

In other respects the criteria listed at the head of the previous section are all valid for resources at these earlier stages of the project.

Subcontractors

In the construction industry the use of subcontractors is a well established method of securing specialised resources which the main contractor does not have available. A particular subcontractor may either be nominated by the client (or his advisers) or may be selected by the main contractor. In either case the selection procedure may well be similar to that employed for the main contractor himself but the timing of the selection is likely to be different. The main argument for the nomination of the subcontractor by the client depends on either his view of the speciality expertise of the subcontractor or the need for the subcontractor to put long-delivery items on order at an early stage.

The whole question of subcontracting is complex and involved but in summary the subcontracting firm, whether nominated or not, is legally employed by the main contractor. The main contractor is responsible for their performance and their payment.

The alternative system which is employed in some industrial spheres is for the specialist skills to be provided in separate main contracts. This imposes a particular demand on the client's management team but this is by no means an insuperable obstacle. On-site coordination of programmes, working areas, site facilities, payment conditions, etc, will be necessary and must be provided by the client. On very large construction sites this type of coordination and control has to be provided anyway so the workability of the system is not in question. Whether it is employed or not will depend on the desire of the client to establish a separate coordination structure or to delegate this responsibility to a main contractor.

Purchasing

Having discussed the problem of bringing labour to bear on the project we must consider the problem of acquiring the necessary materials. The first hurdle is to identify those materials. In general terms this will be obvious from the project data but against each item a full technical description is required. This might be by specification, drawings, supplier's catalogue number or possibly by a statement of the duty which the item is to achieve.

The technical specification must be supplemented by an understanding of when it is required and also by a knowledge of how much money is expected to be spent on the item. There may be other problems such as delivery instructions (is the price ex-works or delivered to site?), insurances, stage payments, guarantees, etc. All these must be resolved before ordering.

It is usual to invite quotations for the items required and clearly it is advantageous to resolve as many of these problems as possible even before quotations are invited. However on receipt of quotations a choice has to be made. As with contractors this is rarely as simple as choosing the lowest price, for many of the other issues could be dominant. Depending on the item a full technical and commercial appraisal is called for particularly if the offer is for an item slightly different from the original specification.

Once ordered, depending on the nature of the item, it may be necessary to verify its quality and that it will be delivered at the stated time. Effort expended on these matters could be costly so judgement is needed to assess its value. It would obviously not be worth while expending any effort on an order for a box of pencils. On the other hand a high degree of attention might be given to a turbine on which quality is of the utmost importance and delivery date vital.

Whatever is decided should be resolved at the purchasing stage as the decision may well affect the price.

On all but the smallest of projects it is worth while having the purchasing undertaken by a specialist as a unique blend of technical knowledge and commercial acumen is required. The handling of a large number of orders at all their different stages through to delivery and payment also demands a specialised system (or battery of systems). The sheer complexity of the operation would be daunting without adequate preparation.

Most construction companies have an established purchasing unit within their organisation and many design organisations are similarly equipped. The advantage of having the construction force do the purchasing is that they are well motivated to avoid delays in delivery if these are going to affect their own operations. On the other hand it may be necessary to order many items before the construction contractor is appointed or certainly before he 'gets steam up' on the job. In either case it is certainly necessary to recognise that the purchasing organisation demands a significant resource and because of its impact on the overall project programme it must be closely coordinated with all the other tasks to be performed.

Conditions of Contract

In considering contracts for labour and purchase of materials the conditions under which the contracts are offered and accepted have a great importance. Some large organisations such as the Property Service Agency and major industrialists have their own standard conditions. There are other generally acceptable standards available from the Institution of Civil Engineers, the Royal Institute of British Architects and other bodies.

To the layman such conditions are formidable because they are so complex. They have been developed to cater for a wide variety of circumstances and to remove misunderstandings or causes of dispute. Thus care is taken to define responsibilities, obligations, ownerships, cancellation provisions and very many similar matters.

Some employers take a standard set of conditions and then modify certain clauses to meet special circumstances. Whatever method is adopted it is important that full agreement should be reached between all parties early in the negotiations so that a smooth path can be achieved. The conditions of contract were never intended to bruise any party or used as a threat. The intention always is to produce a fair and equitable situation in which all parties can operate harmoniously.

CHAPTER 5

Monitoring and Controlling the Action

Part A. Time

At the outset of this chapter we should make it plain that there is an important difference between monitoring and control.

Monitoring is finding out the state of play. It is to do with ascertaining and reporting, whether one is measuring money or time or any other property in which one is interested. It is a vital prerequisite to control but it is a tool needed by control rather than a substitute for it.

Control is taking whatever steps are necessary to vary or alter a pattern of events. It is a positive and active operation the success of which can be judged by subsequent events. Taking decisions in the exercise of control demands sound information which is the result of good monitoring.

If one were to use the terms in relation to a game of cricket the monitoring would be undertaken by the scorer and the control would be exercised by the captain. Both are essential, but the roles are fundamentally different.

Planning

Whether it is actually written down on paper or not much of our lives is planned in the sense that we intend and expect to do certain things approximately in order. This applies to getting up in the morning, making a cup of tea, shaving and going to work just as much as it applies to next summer's holidays. If the activity concerns only one person or a very small unit of people there may be no compelling reason to write down one's intentions. The situation becomes complicated when more people are involved or when one intended activity depends on another. It is at this stage that one begins to write down a programme so that the intentions can be communicated more widely.

Another reason for writing down a programme is to test the validity of assumptions. It is very easy to omit a vital step or to get the sequence slightly wrong if a complicated series of actions is being tackled mentally. If such an error carries a significant penalty it will be seen that there begins to be an incentive for the written programme.

Strictly speaking a programme defines events in the correct sequence and does not become a plan until those events become designated durations and times. This may be splitting hairs because it is often true that the correct sequence of events cannot be determined until the duration of the components is ascertained. In any case for practical purposes the time intervals are usually required. The activity must be monitored against the clock or the calendar and it is unrealistic to suggest that timing is anything less than vitally important.

Drawing up a programme

The amount of detail on a programme can vary enormously as is illustrated by comparing *Figures 19* and *20*. The first programme may have very little detail, with individual steps being broken down into more detail as they are examined more closely. It is not possible to say that any particular level is correct in an absolute sense. The programme is there to serve the job and it is only justified at all if it permits the work to be carried out more effectively and economically.

For example it may suffice initially to reckon that a new project can be designed and built in 30 months. If, when this assumption is placed together with other assumptions the scheme looks attractive, then it will be correct to go into some detail on the programme to (a) verify that the original assumption of 30 months was correct or if not

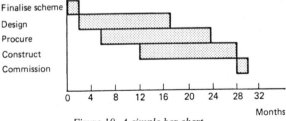

Figure 19. A simple bar chart

what figure should it have been; (b) establish a realistic base against which future action can be monitored and controlled.

For the second of these objectives to be achieved a fair amount of detail must be included. Some effort is necessary to achieve this; thus a more detailed approach should be seen as a second step in the validation procedure, undertaken in enough detail for a sufficiently high degree of confidence to be placed in the results.

Whilst the initial reckoning may be satisfied with a simple bar chart as illustrated in *Figure 19* the more detailed programme indicated above will certainly need a more sophisticated approach. This might still be a bar chart if up to (say) 50 activities are to be included, but for

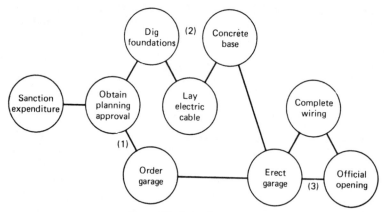

Figure 20. A simple network

Notes: 1. *The garage could be ordered before planning approval if safeguards were provided.*
2. *Since the cable is to be under the concrete it is better done first!*
3. *The 'official opening' does not necessarily have to wait for the electric wiring but the programme displays a firm intention to complete the whole.*

greater detail it is customary for a network to be employed. *Figure 20* illustrates the point with a very simple network showing a number of features of any programme, i.e.

(a) The logic of the intended action has to be thought out, considered and agreed by interested parties.
(b) A reasibably consistent scale of events has to be chosen for clarity.
(c) Interdependencies between parts of the action are established so that they can be dealt with.

What to include in a programme

The difference between a good programme and an indifferent programme is likely to depend on the choice of what items to include. Since each project is different there is no standard to take as a blueprint. It is however possible to make some comments which bear consideration.

Separate entries should perhaps be included for any activity:

(a) which has to be accomplished by an outside or unspecified resource.

(b) which is recognisable as technically difficult.

(c) which depends on results awaited from development or investigation work, particularly if the outcome is likely to alter the programme.

(d) which depends on approval from outside project control (e.g. planning permission, third party participation, etc.)

(e) with a previous history of trouble of any sort.

(f) which is particularly sensitive to pressures which are uncertain, e.g. delicate labour relations, environmental impact, etc.

(g) which closely relates to a focal point in the whole programme, e.g. major road closure for 24 hours, break-ins, temporary shutdowns, etc.

(h) recognised as needing more than one man week's effort.

The list of course is not exhaustive and any new programme will repay careful thought. The list of project steps in *Figure 3* might well be used as a check list to supplement other thoughts.

Intervals of time

Estimates of time required to undertake the various steps will be assembled as the programme is being put together. It is important to recognise some of the principles which apply and some of the values which can be gained from this sort of activity.

(a) As a first step let us recognise that as far as possible people who have to perform tasks should be involved in the decision as to how much time should be included in the plan. This is a method of gaining commitment and providing motivation. Planning should not be a 'backroom' exercise but should be amongst the activity and open for the benefit of all.

(b) The intervals of time used should allow for all the minor or peripheral activities which may be necessary to complete the task but which are not separately included in the plan. For example these might include site visits as preparatory to design activity or extra consultations over some novel aspect of the requirements. The excuses as to why the previous project over-ran might form a useful check list of items which ought this time to be included.

(c) Equally however the time requirement ought to be realistic. There is no point in including every possible contingency and then always completing the task well ahead of forecast. A project plan which is based on unrealistic elements will itself be totally unrealistic. In practice losses and gains never completely balance but it is unnecessary to take steps to make sure that all variations are in one direction.

(d) A record of previous performances on similar tasks will be of value both to the planner and to the person who is responsible for the task. This should not be seen as a method of torture to make life more difficult for the task performer but rather as a measure of efficiency which is necessary so that efficiency can be continuously improved.

Now let us look again at *Figure 20.* It may be a matter of opinion that the completion of the wiring on the garage will only take one day. Maybe the price also says it is only worth one day's work. That however is quite a different thing from saying that it will be done on the next day after the erection is complete. The electrician might be busy on another job or on holiday or still awaiting acceptance of his quotation or waiting for material. To execute the job in the shortest possible time all these problems have to be overcome and the cooperation of the electrician (and his boss) will be essential. This is one good reason for their involvement in the planning.

There is however another aspect with which the planner can help. We mentioned acceptance of quotations and delivery of materials. However cooperative the electrician and his boss might be these two items remain stumbling blocks if they have not been dealt with. There is nothing inherently difficult about dealing with them except that they ought to have been tackled earlier in the action. To deal with these within the system that we are describing it would be desirable to include two further actions in the plan, i.e. 'Select electrician' and 'order electrical materials'. These may have to be dealt with at the same time as the garage is ordered but at least the correct timing can be established once it is realised that the activity is necessary.

If we translate this simple example into a larger industrial situation we envisage that the planner for the project will be working closely with the project manager and assisting him to sort out this type of

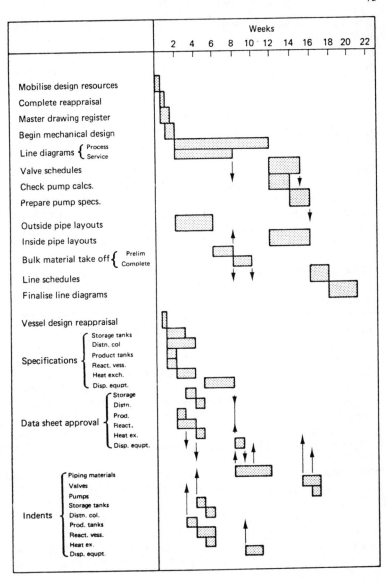

Figure 21. A complex bar chart (part only). Note that information in and out is omitted

problem. Eventually the responsibility for ensuring that a realistic plan is drawn up belongs to the project manager but he will need the help of those who are to execute the component actions and, unless the job is very simple, he will probably need some expert planning assistance.

Advanced Techniques

We have so far tried to keep the examples very simple because it is important to understand the principles. Planning techniques however have made large strides in the past 30 years or so and there are many complete textbooks devoted to the subject. It is not necessary or desirable to go into detail at this point for the connoisseur would do better to read the expert practitioners if he would follow in their footsteps.

One is mindful of Robert Townsend's comment in *Up the Organisation* that 'computer technicians are complicators'. For all its cynicism we do well to remember never to use a computer for the sake of using a computer. The computer is useful when one wishes to do repetitive calculations quickly and accurately. Indeed the availability and reliability of computers have brought many virtually impossible calculations within reach of everyday use.

P.E.R.T. and Critical Path Analysis roll easily off the lips if one has some of the jargon and indeed they and similar techniques have been devised to fill a very real need. No major projects and very few medium-sized jobs are undertaken without detailed planning assisted by a computer. The message of this section is to warn that the understanding of the problem is more important than the use of a particular technique.

The critical path   When the logic of a programme has been agreed and the time of the activities assessed it will be possible to ascertain which particular sequence of events requires the maximum time for completion. This is known as the critical path for any delay on any action in that sequence may be expected to cause delay in the final completion. Of course there may be more than one critical path and there are likely to be several which are near critical and which may become critical during the life of the project.

The first instinct when the critical path is identified will be to check it out to see if some beneficial modification can be introduced. This is fair but the scope will probably be limited and eventually it will have to be accepted that there is a critical path.

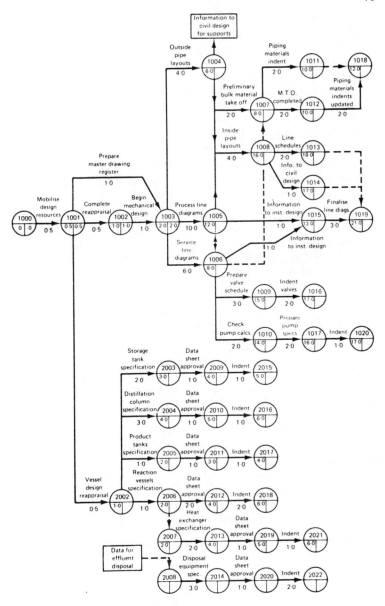

Figure 22. A complex network (part only)

Monitoring the critical path should not basically differ from monitoring any of the rest of the action except that it would be prudent to do it more frequently. It might for example warrant a daily inspection rather than a weekly report. It is not possible to generalise except to draw attention to its importance.

The existence of a critical path implies that activities which are not on the critical path have a degree of flexibility. The jargon word is 'float' and is a measure of time by which an activity is non-critical or simply the excess time which is available compared with the amount needed. The jargon includes terms like 'total float' and 'free float' which are necessary because some activities having float may link in different ways with activities which are on the critical path.

Two general points are worth making about float.

(a) The existence of 'float' on an activity should not lessen the intention to start that activity as soon as it is possible. Projects often over-run because apparently non-critical activities are neglected. The reason for applying significant effort to the planning process is to make sure that *all* activities receive proper attention.

(b) Various theories are sometimes advanced about the declaration of float. It is a comment on human nature that people like to have something up their sleeve—a little something in reserve. This applies as much to the project manager as to the newest draughtsman. The best way of proceeding however is for no one to have anything up his sleeve and have that policy widely known and accepted. The corollary is that no one must be keelhauled just because he has overshot his original prediction. If he is he will make sure it doesn't happen again—by keeping a bit of float up his sleeve!

A reporting system We earlier compared monitoring with keeping the score at cricket and the analogy can be carried one stage further by noting that the scorer also records overs bowled and by whom, how many overs to the new ball, how many wides, no-balls and extras. In fact he runs a comprehensive service offering as many facts as one could wish for either during play or for later analysis. It is similar with project reporting. The purposes of a reporting system will include:

(1) Current performance on every activity which is (or should be) in hand.
(2) Records of delays or problems encountered.
(3) Reports of completion of activities.
(4) Recommendations for updating of the plan where necessary.
(5) Statistical data of numbers employed, size of activity, etc. as will permit performance review at the completion of the job.

It is particularly important to report accurately when tasks are likely to be repetitive either within the one project or from one project to another.

This comprehensive view of what is required from a monitoring system looks good in print but at any instant during the project the emphasis will be on what is happening at that point in time. Inevitably the pressure will be on today's events and so it is important to make sure that the need for longer term interests are not obscured.

Understanding is necessary to make sure that the reporting system is meaningful. For example in steelwork erection it is not enough to be able to report that a certain tonnage has been lifted into position. If the job is incomplete it may not be plumbed or fixed completely. The remaining steel could all be light sections with a higher handling requirement or higher in the structure requiring different equipment or more complex procedures.

In short, numbers are valuable but they must never be divorced from a technical understanding of what they represent or else they will lead to wrong conclusions and be worse than useless.

The other issue to be resolved when considering a reporting system is the circulation list. Not everyone working on a project needs to know everything about all that is going on. The purpose of reporting is that action can be formulated so it must be clear as to how a circulation list can be drawn up. Interactions between different groups, e.g. different design sections, have to be considered but as far as possible keeping circulation lists short is really a protection for those who are omitted and they should be grateful for having less paper to study.

On a large project it may be that different management levels will require different depths of detail. This can usually be arranged without trouble if care is given to the specification of what is required.

Control   At the end of the day the justification for all work done in programming and monitoring is to secure better control on the project. Even the emphasis on statistics and records is secondary to this in that their main purpose is to facilitate the next project. Whether or not there is a next project might well depend on the outcome of the current one.

The problem facing the project manager is always to convert a historical report of what has happened into a prediction as to what will result from past events. If he is satisfied with his prediction in terms of progress on the project he may not need to take any action. If he is not satisfied he will need to take some action which will be aimed to improve the progress.

There are therefore two main areas to consider: (a) Is the prediction as to what will result from the current situation accurate? (b) Is the intended action adequate to meet the circumstances of the case?

The accuracy of the prediction will, to some extent, depend on the sufficiency of the programme and the monitoring of that programme. It must be remembered however that the successful achievement of a particular event in the programme does not, of itself, ensure that the next or any other event will be equally satisfactory. A mistake in the programme could be offset by a stroke of luck with the weather or the ground conditions or just because somebody pulled something extra out of the bag. A certain amount of analysis and understanding is required but all with the objective of understanding what will happen next.

Equally the circumstances of the intended activity may have changed. The predicted resources may no longer be available or a test may have been unsuccessful. The range of possibilities is wide but the question has to be answered for better or for worse—is my programme still valid?

If the diagnosis shows that some corrective action is required a whole new range of possibilities emerges. Does the corrective action affect just one activity or will it have wider repercussions? Will the proposed action have the required effect or should further contingency plans be prepared? Does the programme require major adjustment? Is the monitoring activity sufficient and should the problem have been spotted earlier? Does the proposed action affect any of the other current activities?

It will clearly be desirable to keep life as simple as possible and avoid changes to plans which have already been laid. Nevertheless opportunity rarely knocks more than once and the project manager must have regard to his project objectives in coming to decisions of this type.

Part B. Money

Estimates

Prior to the expenditure of money any view as to how much might be spent forms an estimate. Some people are given to a shrug of the shoulders and some such comment as 'that will cost us 25 millions'. Others are much more cautious in their approach.

We should recognise that the purpose of an estimate is to give information upon which a decision is taken and a course of action agreed. With this in mind even the vaguest of estimates can be very useful provided one understands the basis on which that information is given. A 'back of an envelope' calculation yielding an estimate has its place provided it is recognised for what it is.

The usual way of expressing the accuracy of an estimate is to ascribe plus and minus values to a particular figure. This is really not to be commended for the plus and minus values are so easily forgotten. A range of figures is much more likely to retain its meaning.

As an illustration, if an estimate for a scheme was prepared and gave the answer £3M under conditions which suggested confidence limits of plus or minus 25% the easy way of writing this would be £3M ± 25% or £3M ± 0.75M. The preferred expression would be that the estimate is in the range of £2.25M to £3.75M.

Both ways of expressing this statement mean that one is reasonably confident that the cost will lie in the stated range. 'Reasonably confident' does not mean certainty but it might be held to imply 95% certainty. Thus if one prepared twenty estimates with the accuracy quoted one would expect one of those estimates would turn out to be outside the limits. This statement tells us nothing about how far outside the limits that one estimate will be nor does it tell us which of the twenty estimates it will be. (See also Risk Analysis in Appendix 2).

An estimate can only be as good as the information upon which it is based. If one looks in a newspaper and sees, for example, a car advertised at a certain price one could produce an estimate that was almost certain spot on accurate for one's expectation is that one could go to the dealer and buy the car at the quoted price.

It may be however that the car is not taxed or needs a new tyre or you might choose to fit a radio. If any of these apply the amount of cash necessary will be the purchase price of the car plus something for the extras and at the time of reading the advertisement the plus may not be discernible. This immediately introduces a variability into the estimate and one can recognise that the amount of variability is partially governed by unknown factors and partly by choice.

This simple example serves to illustrate that care is needed in preparing an estimate and even more care is needed in interpreting what that estimate is intended to convey.

In capital estimating for projects the task of the estimator is essentially to ascertain or assume what is required and then to price it accordingly. When he totals up his figures he will know (or must assess) to what extent he has been working on firm data as to what is required and how reliable is his cost information. This will enable him to give his view of the range within which his estimate lies.

Taking these two components in turn the data on which an estimate is to be built will become more and more definitive as the scheme develops. At the outset he may only know that he wants a price for building an industrial plant to produce *Y* tonnes per annum of product *Z*. Later on he may know that the plant is to be located in Scotland and this will mean special arrangements for receiving and storing raw material *M*. He may also know that because the site is not level major excavation is necessary but an intermediate lifting operation can now be done by gravity. He may still not know whether the final treatment stage is to be done by process **A** or process **B** at half the capital cost.

The second component is the cost information about the individual items in the project. The information to confirm or modify the assumptions about price does not emerge until purchasing begins. It is only then that 'firm' prices are really available and in some cases claims and counterclaims may obscure the final price until a long time after the job is completed.

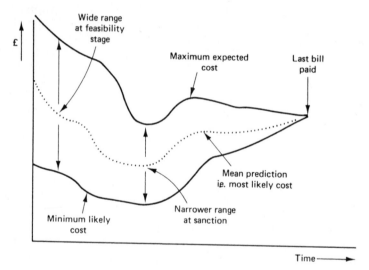

Figure 23. The estimate gradually narrowing

As the shape of the project develops it will become possible to forecast the likely cost. The mid-point view of that cost will also change. This refining process never really stops until the project has been completed, the bills paid and one can say 'that was what it cost' (see *Figure 23*).

Spending

If one sees the estimating process as a gradual narrowing of limits to a final conclusion one must also get into perspective the pattern of spending. It is a matter of experience that for any project which is not subject to major deviation or change of intention the spending pattern

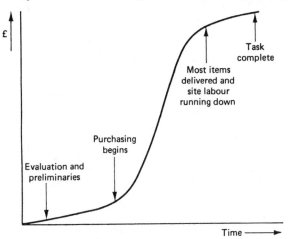

Figure 24. A typical spending curve

takes the form of an S-curve. This is illustrated in *Figure 24*. It will be appreciated that the scale of such a curve could be variable during the life of the project and the illustration is only typical.

Nevertheless superimposition of the view of the accuracy of the estimate onto the S-curve with some of the main events also shown provides an interesting picture (*Figure 25*).

It is worth noting that even when physical completion has been achieved there will still be some small uncertainty about the final cost. This may be because of unresolved claims, unmeasured work or the need for some further expenditure to tidy up or modify what has been achieved. All these may fall within a narrow margin of uncertainty but one or more of these features usually exist.

A further point to note is that the decision to proceed must be taken whilst the uncertainty on the estimate is quite high. It is a fine point of judgement as to how wide an uncertainty can be permitted. Further scheme refinement will improve the accuracy but some spending will be taking place whilst this is going on and the overall effect might well be to increase the total bill. More seriously perhaps the usage of calendar time is being prolonged and this may never be recoverable. In the ultimate a wavering at the point of decision to spend could produce an effect shown in *Figure 26*. This may not be acceptable either.

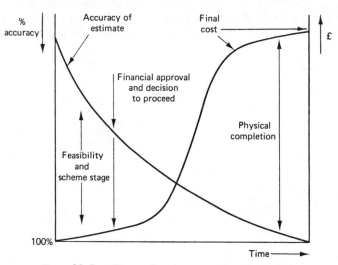

Figure 25. Spending produces a more accurate estimate!

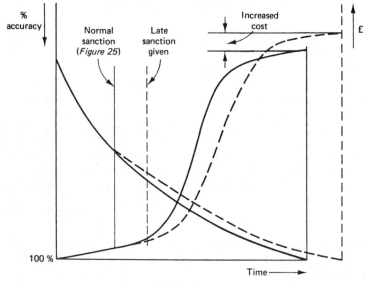

Figure 26. The effect of delay

This further example of the relationship between time and money emphasises that the timing of decision making is very important quite apart from the accuracy of the decision. As far as possible decisions in project work should be made late because the additional information available will tend to improve the accuracy of the decision. There is however an increasing danger of reducing the effectiveness of the decision by such delay.

Many sanctioning bodies would require a risk analysis to be undertaken at this point (see Appendix 2). Even without a formal review of this type the major uncertainties will have to be assessed to appreciate their possible impact on the spending curve. A delay after spending has begun is the worst possible situation to get into and worth strenuous efforts to avoid.

Expenditure and Commitments

Whilst these two words have obviously different meanings there is sometimes confusion in their use. Since they are related to each other this is not surprising but the distinction bears examination.

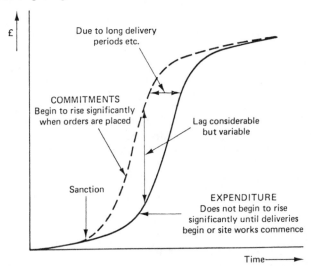

Figure 27. Typical curves of expenditure and commitments

Clearly expenditure refers to the actual passage of money as understood by the bank manager and the amount is specific and clear. If money is paid for a supply or a service that expenditure can be recorded but it tells us nothing as to whether it is an interim payment or whether there is another similar item to be performed and paid for later.

Entering into a commitment of a certain sum of money means that unless action to the contrary is initiated that sum of money will be paid at some time in the future. The time scale is not necessarily specified accurately and there will normally be some uncertainty about the precise amount of money. The intention however is fixed and inescapable.

Because items ordered may have long delivery periods and contracts are let for execution over long periods of time there is always a considerable gap between the expenditure and commitment figures for a particular project. Nevertheless the shape of the spending curve remains the same as illustrated in *Figure 27*. It will be seen however that the gap between expenditure and commitment will vary during the life of the project and its maximum amount will depend on the particular project under consideration.

Monitoring

Whatever may be the confidence limits on any particular item in the estimate the mechanism of monitoring demands that a single figure for each item is used as the yardstick. To establish a good monitoring system for a project the total expenditure needs to be split up into a large number of components and it would be almost unmanageable to have two or three possible target figures for each component.

This leads to the setting up of what may be described as the 'Control Estimate'. It is the target estimate against which the monitoring and later the control is achieved. It does not have to be the same as the sanction estimate or any other document although clearly any divergencies between them must be fully understood by those responsible for the financial management.

An example of such a divergence would be a contingency sum. Contingencies can be built in for a variety of reasons. Examples might include:

(1) A sum of money included for an office block in case a lease which is being negotiated did not conclude successfully.
(2) A sum of money included because the world price of copper was expected to rise and this might produce an additional expense on electrical items which have still to be ordered.
(3) A sum of money included for unforeseen modifications to be specified later, etc.

Another reason for a divergence might be because of a change in scope of the project between the dates of preparation of the two estimates. Yet another reason might be that some of the money sanctioned was to be spent by a different organisation and monitored separately.

To return to the Control Estimate itself, it is worth while establishing this with some care. The most important principle is that as far as possible the estimate shall be framed in the way in which the money will be spent. On the face of it, for the materials part of the estimate this would not appear too difficult and yet it is easy to cause confusion.

To take an example of this problem let us suppose that the project requires six belt conveyors complete with drives. To arrive at the estimate in the first instance the cost of the conveyors may have been deduced from statistics at £Y per metre and the necessary wiring together with the motor taken separately as £X per drive. The resultant estimate might be very accurate and the method is therefore satisfactory.

When the conveyors are actually ordered they will almost certainly be ordered complete with motor and the wiring (excluding provision of the motor) undertaken as part of a separate order. To cope with this the original estimate will not suffice. To convert it into a control estimate some such split as (a) conveyor (b) motor (c) wiring must be made and then when the conveyors are ordered with motors items (a) and (b) can be bracketed together.

Contracts for site labour present a different problem but again should follow the principle that the control estimate should be made up in the way in which the money will be spent. If this is done the problems which arise mainly at the measuring stage can be tackled with some confidence.

Other types of expenditure which involve plant hire, supply and erect contracts, administration or overhead charges, stage payments, bonuses, licence fees etc. can all be divided up into indent or measure and although sometimes highly complex follow the same basic rules.

Once the control estimate has been established the use of it begins in earnest. As far as materials are concerned it should consist of a long list of items which have to be purchased for inclusion in the project. The required system will be one in which alarm bells start ringing if the control estimate is exceeded or if it is exceeded by more than an agreed amount or percentage. The alarm could also sound on underexpenditure but this is not usually necessary or desirable.

If a pump is included in the control estimate at £500 and the purchase is agreed at £475 that would appear to be fine. It is indeed fine provided that both the control estimate and the order value have included (or both excluded) the costs associated with delivery, running tests at makers, provision of test certificates, facilities for inspection, any required fittings or ancillary equipment etc. These items should all be covered adequately in the purchasing procedure but they can play havoc with a control estimate if they are not taken into consideration.

For site labour the details of the system employed will depend on the shape of the contract which governs the expenditure. A sizeable project may include several different methods of payment and each should be recognised as the control estimate is compiled.

The three basic systems are noted below although it should be recognised that their simplicity will be spoiled by contract complications such as bonuses, travelling times, condition money and other adjustments or extras.

Lump sum—If a task is to be completed for an agreed sum that amount of money becomes committed once the contract is begun and will be spent when the bill is paid. Interim payments do not affect the committed amount but will of course be reflected in the expenditure.

Hourly rates (reimbursable)—If the labour is being paid for on the basis of agreed time sheets the monitoring must be done on the basis of periodic reassessments of the time (hours) necessary to complete the task. There will presumably be an initial assessment which forms the basis of the control estimate. As the task proceeds and hours are measured it will not be sufficient to assume that the work remaining is the difference between the original estimate and the achievement to date. The periodic reassessment (probably monthly on a normal job) will result in an updating (increase or decrease) of the view of the commitment. For control purposes it is this prediction of the total commitment which is to be compared with the control estimate to give a view of the state of the project.

Schedules of rates—If the site labour is included within the descriptions of various items in a schedule of rates it is sufficient to monitor it in that form. Each rate which is used may be thought to have a fixed measure but if there is any doubt periodic reassessments would be appropriate as described above.

In each case it will be seen that the monitoring is carried out by comparing the commitment with the control estimate. The actual cash payments are not relevant to this particular consideration because at the time an invoice is presented for payment it will be too late to vary the work which that invoice represents (leaving aside of course any question as to the accuracy of the invoice). The pattern of cash payments is very relevant from the overall financial view of the project but we are here discussing immediate actions which become necessary during the life of the project.

A word of warning is necessary on the question of extra works. However these are paid for they are unlikely to be included in the control estimate unless under some blanket heading such as 'contingency'. During the life of the project some form of extra work is likely to be found to be necessary and the problem is to bring this into the monitoring system at the earliest possible moment. Ideally a payment method and a price should be agreed before each item of extra work is committed. This will not always be possible and insistence on the point would be too restrictive. A sharp look-out must therefore be kept for this source of variance. In some projects the extra items amount to very considerable sums of money and it is vital that they should be brought into the system at the earliest possible time.

One of the best ways of doing this is to use a subdivision of the control estimate which might be labelled 'contingency' or 'extra works' or given a more specific title. When the extra works are recognised they are immediately estimated in like manner to the original control estimate. The amount allocated for extra works in the original control estimate is then reduced by the amount of the new estimate. Monitoring of the extra work can proceed as if it was part of the original estimate.

This procedure can be repeated up to the time when the original allowance for extra works is exhausted. At that point it is probably wise to re-estimate what further sum will be required to cover 'extra works' and establish a supplementary control estimate for monitoring purposes. This action also permits the sanctioning authority to become acquainted with the possible overexpenditure.

In some cases with extra works the amount of money required will not be known accurately at the time the adjustment is made to the control estimate. There should however be no delay in making the adjustment as it will affect the accuracy of the monitoring process.

Converting Monitoring to Action

If monitoring is properly carried out it will provide an accurate forecast of the Anticipated Final Cost at any point in time during the life of the project. It will be a matter of choice whether this forecast is reviewed and updated monthly or at any other frequency but regard should be paid to the review frequency early in the life of the project.

If the monitoring shows that the control estimate is going to be exceeded by a significant amount or if the trend is rising the project manager will need to make some quick decisions. Usually the choice is simple but hard for it customarily lies between accepting the prediction of overexpenditure or cutting back on the scope of the project. Both will be unpalatable courses of action but once again we see the emergence of the conflict between time, money and quality.

No abstract advice can be given as to which course is right for it will depend upon the circumstances of the individual project. The best advice that can be given is that the problem should be tackled quickly and positively. If the figures have been assembled accurately the problem is real and it will not go away. Delay in tackling the situation will narrow the scope for action for already there will have been further movement to convert commitments to expenditure.

There is a significant point to be looked for in reviewing costs. If overexpenditure on an item is reported through the system we are using it will be probably too late to take any corrective action on that particular item. The important question however is as to whether the *reason* for that overexpenditure is likely to apply to other items in which there is still some freedom of action. Thus if the reason is a blanket increase in the basic price of an essential material such as steel or cement it may be inferred that other items will be affected for the same reason and conclusions reached accordingly. On the other hand if the reason for overexpenditure was an isolated one the implications may not affect the rest of the project.

Incidentally the obvious need for urgent and decisive action involving the fundamental issues on which the project was based, confirm the desirability of having the management of the project under one control. The project manager may have limits set upon his authority both with regard to additional expenditure and modification of the scope but it would be intolerable if the lines of communication at this crucial stage were diffuse or duplicated.

Part C. Quality

As a general rule (although obviously not without exceptions) the employment of the appropriate professional, technician or craftsman for a particular task will go a long way towards ensuring adequate quality. The point is that by tradition people qualify into their area of expertise by tests of ability rather than by tests of speed or by the price they charge. From this it may reasonably be argued that bringing the right expertise to bear on a problem is a major task for the project manager.

This takes us right back however to the dilemma which we recognised early on in that we don't always want a Rolls-Royce job. The desire to get a high level of quality must be tempered by considerations of time and cost so it is a question of striking a balance which is most suitable for the particular project in hand.

At the risk of repetition (but it is fundamentally important) the right level of quality has to be established by or by reference to the objectives of the project. This is one very good reason why clarity of objectives is so important.

Setting the Standard

The particular quality required, once determined, has then to be defined to others either to implement or to use in relation to their own work. The extended use of British Standard specifications (or the continental or Americal equivalents) has facilitated this problem to a large extent but such standards frequently leave room for choice within those standards and there is always the choice as to whether or not they are appropriate anyway.

In some instances specifications have to be specially written. This would be likely if there were a degree of novelty or secrecy about the technique or material being used. It might also be appropriate if there were a particular assembly instruction or operational procedure required to produce the effect. Such special requirements need to be written down carefully for checking and faithful interpretation.

Of course the engineer's traditional method of communication is a drawing and on any major project there will be many of them. Drawings however are only as good as the effort which goes into them and essentially express the ideas of the person who prepared or approved them. Although a drawing which has been approved should be implemented faithfully, it is necessary to ensure, before approval, that it does comply with the general requirements of the project.

Quality Assurance

Once the level or the standard of quality for any part of the project has been determined and described at the Purchasing stage, the next problem is to ensure that it happens. To ensure that it happens costs money and that amount of money is variable so some sort of judgement is necessary to determine what it is worth to make sure that a given quality is achieved. This is likely to vary over different parts of the project. As we saw in Chapter 4 it would be unreasonable to take the same amount of trouble over a box of pencils as over a turbine because the penalties for faulty work (or late delivery) are vastly different.

The techniques available for quality assurance vary considerably according to the problem being considered. The possibilities include:

(a) Chemical analysis, e.g. sand, cement, steel etc.
(b) Physical analysis, e.g. size gradings of aggregates, strength tests on concrete cubes etc.
(c) Dimensional checks, e.g. tolerances and fits, clearances and size and shape generally.
(d) Condition checks, e.g. appearance, checks on completeness, stage checks such as weld preparation or reinforcement positioning.
(e) Radiographic tests, e.g. welds.
(f) Performance tests, e.g. pressure tests on vessels or pipes, running tests on motors.
(g) Special tests for special purposes, e.g. loading tests on driven piles, noise tests etc.

Frequently testing to ensure quality is included within British Standard specifications and in some cases there are Standard Specifications devoted to the whole question of testing and quality assurance. Certainly the tests which are likely to be required should be determined at the design stage because (a) they form an important part of the design philosophy—the whole purpose of design is to achieve an effect (i.e. an objective); (b) tests called up at a late stage are damaging both to the programme and the budget.

The question of who should do the tests is also important. We referred to this under 'Commissioning' in Chapter 2. In approaching this problem we should remember that the need for Quality Assurance starts very early in the implementation stage of the project and in duration probably exceeds the commissioning period because of the tail of dealing with spares, modifications or replacements. The sponsor of the project may elect to deal with quality assurance himself or alternatively could delegate all or part of the work to the design, purchasing or construction organisations.

Alternatively there are firms who specialise in quality assurance and who would therefore offer the sponsor an independent service to his requirements. In any of these cases it is worth repeating that it is important to determine in advance what level of effort is justified for any given project.

CHAPTER 6

People

Throughout the previous chapters we have been considering actions and how those actions can contribute to the success of a project. These actions are performed by people whether they be located at the site of the construction, the design office, the supplier's works or elsewhere. The vast majority of the money spent on a project finds its way into the pockets of people who do some work specifically for the achievement of the project. The value of the raw material (steel, cement, aggregate) or off the shelf items (perhaps small pumps, switches, kerbs etc.) is comparatively small.

The significance of this is that the ultimate success of the project depends on a lot of people doing a good quality job at the right time. Organisation is aimed at this effect but since people have freedom of choice they can choose to cooperate or not as they will.

Managing the Managers

The most difficult problem facing the Project Manager may well be the difficulty of orientating senior people to the requirements of the project. Depending upon the particular organisation which is established some of these people will be from the same 'stable' as the Project Manager. In any case they will have significant relevant experience and have reached their present status through success in their own particular field.

The Project Manager may be denied direct executive authority over such people and yet he has the problem of securing their contribution to the project—sometimes at variance with their own ideas as to what ought to be done.

The strategy which the Project Manager needs to adopt will clearly vary with the individuals involved. The strategy must be based however on the objectives of the project and the Project Manager's determination to achieve those objectives. He may compare his situation with a general conducting a battle or a grand master playing chess. In reality the project manager needs to have a bit of both approaches. The 'opposition' may be less tangible in the sense that they would not wish to be labelled as 'opposition'. Nevertheless they are every bit as difficult to overcome with complete success.

Construction Phase

We have seen already how the expenditure curves look on a typical project (*Figure 24* etc.). These are very steep shortly before the completion of the project. The steeper the better from a cash-flow point of view. The steepness of the curve is a function of the number of people contributing at that particular point in time. It follows that the risk of delay through non-performance is greater towards the end of the project.

Traditionally this is the period when large numbers of craftsmen and unskilled labour are employed on site. This feature has thrown considerable emphasis on problems which have been experienced in certain industries in labour relations and in union activities. This is not the place to discuss the problem at length for there are complex issues involved. From the project point of view however let us note some of the ingredients of the situation.

(a) The large number of people who contribute to the construction activity have probably been brought together specifically for the job in hand. It is likely that the majority of them have no previous experience of working with the others on the site. This is true even of men employed by the same contractor for on a large job many of them will have been recruited specifically for the job in hand.

(b) There are likely to be several different management teams belonging to contractors and subcontractors. They will provide several different management styles and have varying degrees of capability. They will react differently to the problems encountered on the site and this of itself causes several different manager/man relationships.

(c) Tasks in the construction industry are always of limited duration and traditionally there has been little interest in getting long-term solutions to problems unless the short-term is first made satisfactory. This situation has changed somewhat with more enlightened employers and union attitudes. There remains however the basic fact that, on any job on which there is a problem, there is a limited amount of time in which that problem can be sorted out.

(d) The fact that the owner or sponsor of a project has already made a significant investment of capital when the construction phase begins leads to a potential 'ransom' situation. There is no income from the investment until the task is finished and thus the threat of delay, e.g. by strike can have very serious short term consequences for the owner. Under these circumstances it is understandable that an owner will sometimes be willing to pay a premium to avoid delay. To what extent this is ethical, wise or prudent must be judged in a particular situation. The reality of the pressure cannot be ignored.

(e) A lot of construction work is exposed to vandalism and theft unless very tight security is practised. Unless the site is well controlled this feature leads to suspicion and mistrust. In particular selective damage or theft is another way of causing delay which could be a greater motive than the value of stolen goods.

(f) Some areas of construction activity have developed a reputation for militant political activity. This may stem partly from the casual nature of so much of the employment. It is fostered and encouraged by the magnification effect of any action which affects the cost or completion date of the project. Thus political motivators find that they are moving in a highly responsive environment and they thrive on it.

The reality of these ingredients and the skill which is necessary to deal with these problems highlights the importance of construction site management. There is no doubt that experience is invaluable and so a specialist branch of the management arena has grown up. No project of any magnitude should disregard the need for this type of skill in dealing with labour problems on construction sites.

The need to keep this important area under control must affect earlier project stages and it is vital that action to establish control is taken sufficiently early.

Earlier Phases

We should however look at a broader canvas than just the construction worker. Similar basic problems face the manager of resources used earlier in the project. In this case the problem may not be so dramatic as a work or strike choice but it is none the less crucial. This stems from the fact that work contributed early in the project has usually a much more radical effect on matters which affect the profitability (*Figures 28* and *29*).

We cannot fairly say that people employed at earlier project stages are noted for their militance although recent years have seen the growth of organised pressure groups in drawing offices. There are also those who would seek to unionise research and development groups.

These groups have not got the same weapon as the construction worker in that their activity takes place substantially before the significant portion of the capital investment. The 'ransom' factor is therefore different, having more of a technological content than sheer cash.

Reference to *Figure 16* will demonstrate that the problem is one of technical management. The man working on the project may have the

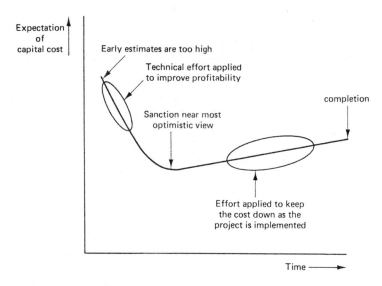

Figure 28. Cost predictions vary with time

requisite technical skill but how can one ensure that he is working to the maximum of his capacity. The problem exists both in terms of the rate of output and also in terms of the quality of the contribution.

Since much of the early work is creative the handling of the problems in this area needs a great deal of sympathy and understanding. Typical attitudes which have to be dealt with are of unwillingness to be pushed or chased into making arbitrary decisions or wishing to search for perfection even when it is obvious that the law of diminishing returns is applicable.

We are saying that the way in which every person contributes to a project is important in that it could affect the timing of the project or the quality of the project. If either is adversely affected the cost is

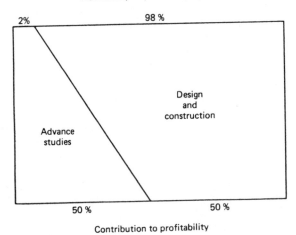

Figure 29. Early effort contributes to profitability

sure to suffer. Part of the management job and probably the most important part is to establish the environment in which all who contribute can make their efforts most effectively.

Behavioural Science

It was inevitable that the complex patterns of human behaviour would sooner or later come under the scrutiny of the scientific approach. The analytical and searching approach which has been at the root of the enormous technological progress of this century has been focused on the people who are themselves instigators of this progress.

So a new science has grown up and with its development a great deal of jargon, very many hypotheses, and a wealth of literature. In no way should we condemn the work that has been done but we must recognise that it is a field in which a great deal of specialisation has taken place. As usual with a subject which has become specialised, the person who wishes to use the results of the specialisation but without becoming a specialist has got a real problem.

Managers will vary on the extent to which they wish to study (or need to study) behavioural science but all should appreciate the benefits which could follow their understanding. The choice is then open to read more than these few shallow sentences which follow.

In essence behavioural science is to do with the reactions of people to the circumstances which surround them. The argument quite simply is that if the reactions can be predicted then the circumstances can be so arranged to produce the reaction which is required.

The benevolent view is obviously that such a study can lead to social improvements, better working conditions and a better environment for people at all levels of the organisation. If one takes a more material view the study could lead to improved performance in terms of either quality or volume or both.

In the extreme however there is lurking in the background the thought that knowledge is power and a knowledge of how people behave gives power over those people. In the ultimate this is somewhat frightening and leads to the conclusion that if behavioural science is to be studied at all it should be studied and understood as widely as possible.

This expression of concern is only a personal view for there can be no doubt that the origins of behavioural science spring from a compassionate desire to improve the human lot. After all increased productivity should give higher reward and more leisure as well as increased satisfaction to all concerned. Behavioural science of itself is not particularly moral or immoral. This question depends on the use to which the findings are directed.

Herzberg made a major contribution to the thinking of behavioural scientists in his book *Work and the Nature of Man*. As a result of observations he identified two broad groups of factors which affect the individual worker.

The first group which he called 'Maintenance Factors' are necessary to be present in order to achieve normal job satisfaction and hence normal performance. The absence of these factors leads to dissatisfaction but their presence does not lead to any positive motivation. They are thus commonly called Hygiene factors and include:

> Company policy and administration
> Supervision (technical)
> Money
> Interpersonal relations
> Working conditions
> Status
> Job security
> Personal life

These, as it were, represent the background of the arena in which a person is to operate. It must provide him with the necessary environment or else the job begins to suffer. They are factors which need attending to as a minimum of activity.

The second group were called 'Motivation Factors' and positively amount to carrots which will produce greater satisfaction and higher performance. The absence of these factors does not actively create dissatisfaction. The motivators include:

Achievement
Recognition of achievement
Work itself
Responsibility
Advancement
Possibility of growth

In advancing this theory Herzberg had built on the 'hierarchy of needs' which Maslow had put forward. This theory had established

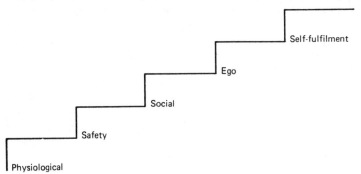

Figure 30. Maslow's 'hierarchy of needs'

various levels of needs and suggested that motivation was always based on a need being satisfied and used as a stepping stone to the achievement of the next higher need. These needs are illustrated in *Figure 30.*

If a person is continuously striving to ascend this ladder (as they will be attempting to do by their own efforts), the manager's job must be to encourage them. As they succeed their performance will improve and, in a managed situation, everyone will benefit.

The manager who seeks to look after his people as well as get a particular job done must have an appreciation of all these factors. They affect the way people respond both in the short and long term. To some, the appreciation will be almost instinctive but others may need to adjust their thinking in the light of this approach. All will come to realise that an understanding represents a major contribution to the successful management of a project.

Selection for Project Management

We may have established that the project manager ought really to be something of a superman. Not only will he be gifted with all the superlatives of the qualities available but he needs to be fully aware of all events in the past, present and future. Most of our projects will have to be managed with mere humans so we ought to pay attention to the questions of selection and training.

The job of managing a project which will involve responsibility for spending a great deal of money is unlikely to be entrusted to anyone who has not previously had experience of what goes on in project work. It all happens too quickly and the decisions are too far-reaching and final for the job to be entrusted to a man without experience. One is always building bigger or faster or better or at least under different conditions so each job has a degree of novelty. Exactly similar experience cannot necessarily be arranged but at least it should be relevant and helpful. Not only will the project manager be glad to be able to use his experience but also others around him will have much more confidence in the success of the venture.

Then, we may ask, is the experience to be preferred in the construction field, in design work or possibly in development work? Or is there some other experience which is relevant? It is unwise to be categoric or someone will quote the project which was completely successful but managed by a genius from some other environment. Nevertheless the following points should be borne in mind.

(1) The greatest contribution to the saving of capital on a project can be made by establishing the technical basis for the project at the earliest possible stage.

(2) The greatest disruption to the planned course of a project can occur shortly before planned completion when the project construction is in full swing.

(3) Most modern projects of any size embrace a wide range of technology which must be integrated into a viable whole.

These suggest a minimum competence, through experience, embracing high technical skill with at least a sympathy if not an understanding of related technologies. A construction experience is also implied because it is here more than anywhere else that the fruits of earlier decisions can first be seen. The volume of construction work also provides a situation where all the technologies involved are brought together in the focal point which is the project.

It goes without saying that one believes the project manager should be an engineer. There might be a degree of prejudice in this view but the objective of the project manager is to get the project built. An engineer is someone who gets things done! The formal training and usual experience pattern of an engineer is aimed at making him a better engineer. In some cases it will also make him a better Project Manager.

Training for Project Management is also a subject which is tackled by a number of educational institutions. Generally the approach is through a short course concentrating on one or more aspects of the subject.

We have mentioned in earlier chapters many aspects of the management problem including the wider understanding of management itself. Organisational Theory, Behavioural Science, Planning, Capital Cost Estimating and Control, Communications, Project Assessments etc. are all topics which can be taken on one side and studied in depth. They all contribute to the total subject of Project Management.

A manager or a potential manager who has had some experience of project work will probably feel that his practical experience equips him well enough to deal with some of these topics. In his case then it is most appropriate that the short course specialising in a limited field is chosen. Not only can a gap be filled in this way but also it provides an opportunity of keeping up to date with current techniques or current thinking.

We can find further approval for the short specialised course concept by noting that it provides an opportunity to stand aside and develop a perspective view of the whole scene. This contributes an important aspect to the achievement of the balanced view which is the project manager's prime essential.

As an example of this point consider a man with a typical engineering attitude which lays great stress on the capital cost of a job. This may have resulted from a background of cost responsibility which demanded tight estimates and fulfilment of those estimates. If a thorough understanding of Project Assessments is superimposed on that background it will be easier for the man concerned to make balanced decisions which sometimes demand that the capital cost estimate is exceeded. We are not denying the importance of cash budgetting but we are opening the gate to a wider understanding of true value.

CHAPTER 7

Looking Forward

A volume such as this would be incomplete without a forward look at the way in which the arena might change. This is appropriate not in the sense that a tipster will predict the outcome of the next race but in the sense that one should be prepared for changes which may be expected. There is also a humble aspiration that, by drawing attention to a number of weaknesses, debate may be encouraged and remedies devised. In some cases too remedies are on the way but have not yet gathered momentum to have become widely accepted.

With these varying thoughts in mind we will remove the cover from the crystal ball and take a look at a few topics.

Total Technology

On the face of it this might be a jargon phrase but it represents an approach to a problem which we have referred to several times. This is the rapidly increasing technology which is required to become encompassed within the boundaries of even the most modest of projects.

The Science Research Council has become interested in the problem of encouraging suitable training for individuals to tackle jobs which demand a synthesis of varying technologies. To identify a common basis for action the 1973 booklet on Total Technology attempted a definition.

> The practice of engineering comprises research, development, design, production, marketing, and operation of plant. In addition the service and construction industries require a special emphasis on planning and operations management. The parts of this continuum of functions merge into each other with ill-defined boundaries and any one has a marked influence on and inter-relationship with the others. Success can only be achieved with well-balanced syntheses of all functions. Total Technology is the name given to cover this wide spectrum of functions in the practice of engineering coupled with the skills required for welding them together.

The definition is somewhat long-winded but it is conscious of a desire not to tie down the thinking on Total Technology to too narrow a channel. If we are guilty of such narrowing in our consideration of project work it is because we are trying to be too specific. In fact training for Total Technology is very appropriate as training for Project Management. Indeed the practice of Project Management as we have described it in this volume is very much an exercise in Total Technology.

Our hope for the future is that there will be a growing understanding of the desperate need for technical management when projects are being contemplated. The growth of the concept of Total Technology should encourage the engineering professions generally to equip themselves more completely for the tasks which lie ahead. Hopefully the result will be a growth in confidence that will permit full advantage to be taken of all the advances in technology which surely must be made.

The Sponsor's Interest

The management of projects is becoming such an important and complex problem that the sponsor is faced with a difficult choice. Either he will manage the project himself or with his own staff or if he chooses, firms of consultants or contractors will undertake management roles for him.

There is a lot to commend the use of his own staff at least for the management role. In so far as the major decisions will affect programmes, budgets and standards to varying degrees, the choices are often best evaluated by the sponsor himself. If he retains this right he can at least be satisfied that all his experience and all his wishes have been considered in arriving at the solution.

The difficulty arises when the sponsor does not have the necessary skill and capacity within his organisation. He may be tempted to undertake the task using as Project Manager someone who does not have the necessary aptitudes. Alternatively he delegates some of his responsibilities to others have have a different part to play in the project.

This situation has led to the growth of organisations whose sole responsibility is to manage the project. They are divorced from design and construction although it is likely that some of the staff will have considerable skill and experience in both fields. This freedom from other tasks on the project gives the manager the opportunity to concentrate on management aspects and to promote the sponsor's interest with undivided effort.

Such a solution has particular value if the project is complex or specialist or if it is required in a short time. We have seen that project management is a demanding task, rapidly becoming a specialisation in its own right. We can appreciate that techniques are likely to become more sophisticated. It is therefore a considerable advantage to be able to combine a safeguard of the sponsor's interest with control of his project.

For the future it is likely that the use of such specialists will grow. It is true that they represent an increase in the overhead which the project incurs. It is claimed however on projects which have used this method that the additional cost has been more than offset by savings on other aspects of the project.

An American Approach

We must here refer to the American practice which comes under the general heading of Construction Management. This appears to stem from the attitude that gave us PERT and other most complex planning achievements. These have demonstrated beyond doubt that even the most complex operations can be achieved in an incredibly short time if sufficient attention is paid to detail.

These achievements have led to particular emphasis being placed on the time factor of projects and some very short time scales have been recorded. The emphasis on time has also been felt in organisational relationships to the extent that particular systems have been developed which are directed primarily at the achievement of short time scales.

Construction Management is a generic name for an approach which has been developed in recent years. It is based on the premise that it is the construction phase of a project which offers greatest scope for telescoping the time scale. Time gained will inevitably be accompanied by the saving of money and of course early achievement of a project means an earlier start to the income from its use.

This great emphasis on the construction aspects leads to the suggestion that 'construction' attitudes and approaches should be dominant right from the moment the existence of the project is acknowledged. In fact the management of the project should be in the hands of a construction orientated manager from its inception. This is an organisational requirement. Techniques and tasks are subservient to it but all contribute to the early completion of the project with the consequent financial advantages.

The American scene provides a good background to this attitude being adopted successfully. There is a different attitude to the taking of technological risks which results in more extremes of advantage or disadvantage being gained. If a greater risk is accepted it is better to get on with it quickly so that the corrective action necessary can also be set in motion more quickly. Alternatively if it is accepted that (say) one project in four will be a commercial failure it is better to get on quickly with all four so that the three winners will pay dividends early. These attitudes are part of the scene in which 'hurry' is the operative word.

American attitudes are also more flexible and there is probably a more likely chance of unusual requirements being achieved. This is a point which is well worth pondering when faced by excessive tradition.

Finally it is also generally true that American delivery periods are favourable. Thus a project timetable does not have to be dominated by massive periods of waiting.

The improvement in performance which is claimed by the Construction Management approach however is not offered solely by contractors. Specialised firms as noted in the last section offer the necessary services as also do consultants whose original background was a design activity.

In all approaches however the common feature appears to be that the intended actions which are necessary for the achievement of the project are initially circumscribed most carefully by budgets of time and money. All subsequent action is then monitored and controlled within this envelope.

This is truly what Project Management is all about and we may expect the American approach to be studied, welcomed and adapted for U.K. conditions within a short space of time.

Objectives

There is little doubt that the sophisticated techniques which have been developed for time and cost control will become more widely used. As they become more widely used further techniques will be developed. There is little doubt that this side of the management problem will receive a great deal of attention.

It must however be balanced by holding the standards of the project in proper regard. It is like the swing of the pendulum. In the past we may have been obsessed with quality without sufficient regard for time and money. This is now being corrected but we must be sure that the swing is not excessive and if possible damped down.

Holding the balance of the project can only be done by getting the objectives into proper perspective. In the past this has frequently been neglected but we may now hope (and believe) that there will be a fresh understanding of the importance of this aspect.

In particular we can expect the clarification of objectives to have an impact on the early stages of project work. Many projects suffer neglect in their young life due to pressure of other work, lack of a sense of urgency and perhaps uncertainty about what should be done next. Clarification of the project objectives should make a major contribution to bringing the project into being at the right time and supported by the right amount of preparation work.

Environmental Issues

We referred to Environmental Impact Assessments in Chapter 2 and mentioned the increasing awareness of the importance of conservation in its widest sense. There is little doubt that this will have a great influence generally on future projects.

Whatever our degree of enthusiasm for conservation matters we should welcome a more active consideration of those matters which affect the quality of human life. The difficulty however will be to keep a balanced and perspective view of the situation. The experience on some recent projects has been that minority pressure groups, having put forward a point of view which has been incorporated into the project decisions, have still exercised their pressure. Whilst a 'never say die' attitude may be appropriate in warfare it is not helpful to those who are executing projects which are designed to achieve acknowledged benefits.

Perhaps this is a fitting note on which to draw this survey to a conclusion. We have emphasised that project management is not easy but it is a challenge which must be accepted and can be enjoyed.

The Time Value of Money

The Time Value of Money

With high interest rates the relationship of time and money is increasingly important. No longer is it simply a question of calculating how much money is required and ascertaining that funds are available. It is now vitally necessary to relate the cash flow analysis to a time scale in order to be able to portray an accurate picture of what is required.

To quote an oversimplified case to demonstrate the problem let us assume that a project costs £1M and the investment is spread evenly over a 12 month period. Let us further assume that there is an expected profit which amounts to £250 000 per annum each year after completion. We ignore inflation and variations in profitability and may say that the simple pay back period is four years after initial operation and five years after job start.

Now let us superimpose the fact that to finance the project money has to be borrowed at 12% per annum (a low figure!)

To ease the calculation we shall also assume (incorrectly) that the calculations are made on the outstanding debt at mid-year.

Year 1	Capital expended	£1 000 000
	Interest (12% on £500 000)	60 000
	Balance at year end	− 1 060 000
Year 2	Operating profit	+ 250 000
	Interest (12% on £935 000)	− 112 200
	Balance at year end	− 922 200
Year 3	Operating profit	+ 250 000
	Interest (12% on £797 200)	− 95 664
	Balance at year end	− 767 864
Year 4	Operating Profit	+ 250 000
	Interest (12% on £642 864)	− 77 143
	Balance at year end	− 595 007

Year 5	Operating profit	+	250 000
	Interest (12% on £470 007)	−	56 401
	Balance at year end	−	401 408

Thus the developer instead of having paid off his debt at the end of the year 5 will still owe £400 000—a nasty shock!

An even more dramatic result can be shown if one assumes that there was a 6 month delay in completing the project. Let us assume that the effect of this is to halve the profits in the first year but that after that no change is made to the circumstances. The sum now looks like this:

Year 1	Capital expended		£1 000 000
	Interest (12% on £500 000)		60 000
	Balance at year end	−	1 060 000
Year 2	Operating profit	+	125 000
	Interest (12% on £997 500)	−	119 700
	Balance at year end	−	1 054 700
Year 3	Operating profit	+	250 000
	Interest (12% on £929 700)	−	111 564
	Balance at year end	−	916 264
Year 4	Operating profit	+	250 000
	Interest (12% on £791 264)	−	94 952
	Balance at year end	−	761 216
Year 5	Operating profit	+	250 000
	Interest (12% on £636 216)	−	76 346
	Balance at year end	−	587 562

The increase in the debt at the end of year 5 is not only greater by £125 000 because of the loss of profit in year 2 but also a further £61 000 which are additional interest charges.

All the calculations are simplified but serve to illustrate a very important principle. It will be apparent that the principle increases in importance as interest rates rise and in periods of high inflation.

Discounted Cash Flow

Before the accountants swamp us with figures let us take time off to look at another phrase which is used to help bring this time and money relationship under control. Essentially the examples worked out above are discounted cash flow exercises but a rather more elegant method of handling the figures is necessary.

The principle of compound interest will be familiar and can be expressed by the term

$$T = P(1+r)^n$$

where T = the total sum of money at the end of the calculation,
  P = the initial amount of money (principal),
  r = rate of interest per annum (or other period),
  n = number of years (or other period).

If we go back to the first example quoted and consider the profit made in year 2 (£250 000) we might see other ways of making that profit. Using the compound interest formula and the same rate of interest (12%) we can say

$$P = \frac{T}{(1+r)^n} \quad \text{or} \quad \frac{£250\ 000}{(1+0.12)^2} = £199\ 298$$

Thus if we invested £199 298 in a fund yielding 12% per annum, at the end of two years we should have our £250 000. Another way of saying the same thing is to say that the 'present worth' of that £250 000 is £199 298. Having £199 298 today is of equal benefit as having £250 000 in two years time.

Similarly the profit made in year 3 (250 000) has a present worth of

$$P = \frac{£250\ 000}{(1+0.12)^3} = £177\ 948$$

Remember that these are simplifications. We should get a slightly different picture if we assumed that accounts were made up every six months and an interest rate of 6% per 6 month period was used. Similarly it would be different if monthly or daily calculations were used.

Furthermore we have assumed that we can borrow money at the same rate as we could achieve from a deposit! Nevertheless let us proceed.

As far as the £1 000 000 investment is concerned we can also find the present worth of our outgoings. If we assume that the money is spent on the first day of the exercise the £1 000 000 clearly has a present worth of £1 000 000. If, on the other hand, we assume that all the bills are paid at the end of the first year the present worth of the investment

$$P = \frac{1\ 000\ 000}{1.12} = £892\ 857.$$

In truth of course the figure will lie somewhere between these two extremes.

However complex the arithmetic may become the principle which we are looking at is important. It is that future incomes or expenditures can be compared with each other if we bring them to a common basis which is their 'present worth'. The process of bringing these sums of money to their present worth is known as discounting. Since all the sums we are now dealing with are discounted (i.e. back to their present worth) we are building up a picture of the discounted cash flow for the project.

The complexities of the arithmetic are even greater than we have indicated. If the technique is to be of value to us we would wish to consider in the example chosen such variables as different sales forecast, different selling price, different operating costs etc. All this would be very laborious so it is customary to commit all but the simplest of cases to a computer programme. If this is done carefully and all unwarranted assumptions taken out we now have a valuable method of studying the effect of variables in the future—an invaluable aid to forecasting.

Let us note in passing however that discount tables are available to give values of the expression $1/(1 + r)^n$ for different values of r and n. These tables enable one to achieve simple calculations quite quickly.

The main use of the technique will be in evaluation of different schemes or of different methods of executing the same scheme. The procedure commonly adopted would be aimed at finding r—the interest rate. To do this it is assumed that the present worth of the expenditure equals the present worth of the incomes. The period of time chosen might be the life of the asset, the expected life of the market or some arbitrary period which represents the maximum time which the business would allow. With n fixed and the actual sums of money assumed fixed the calculation can be completed.

The rate of interest at which the present worth of expenditure equals the present worth of incomes is defined as the interest rate of return. A high interest rate of return is clearly advantageous.

Alternatively projects may be compared solely on the basis of their net present worths. Again assuming a fixed period of time the present worths can be graphed against various interest rates. If a business demands a minimum interest rate it is possible to compare propositions at that and higher interest rates on the basis of their present worths at those rates.

Once a project has been chosen and started it is unlikely that a complex calculation will be needed for project decisions. Nevertheless an understanding of the principles involved will help.

APPENDIX 2

Risk Analysis

Risk analysis is the name given to a technique which seeks to evaluate and compare the various issues which must be considered at the assessment stage. We will attempt no more than a broad understanding of the need of the technique and the way in which it can help for in practice it becomes highly complicated and specialised for the particular type of project in mind.

Whether it should be used or not for any particular project will depend on the degree of complication and how marginal the issues appear to be.

In the simplest possible situation we might consider the tossing of a coin when that is used as the basis for a decision. It is very appropriate for two football teams when the captains wish to decide which of them shall choose which way to play. They both have equal chances—it is a 50/50 situation and eminently fair. If the coin is to be spun once there is a 50% chance of the guesser being right—no more no less. If however the coin was spun twice and he said 'heads' both times he has a 50% chance of being right once and a 25% chance of being right both times. Even so he may still be disappointed both times and he has no guarantee of success.

If the coin were spun 10 times we know from our experience that he would indeed be unlucky if he was never right but we equally recognise that he has no guarantee of success. If we were to express this mathematically we would say that there was slightly less than a 0.1% chance of him being so unlucky!

The point of this is to show that it is possible to ascribe values to risks so that they can be expressed in meaningful terms. Once a value can be assigned it is possible to make comparisons between factors which are otherwise as incomparable as apples and pears. Furthermore it is possible, having quantified risk, to decide whether a certain level of risk is acceptable or not. Turning specifically to an industral project there are many areas of risk which might be considered. These would normally include but not necessarily be limited to:

> Amount of capital required.
> Cost of start up.
> Operating costs (many components, e.g. labour, services, overheads).
> Raw materials costs.
> Product selling price.
> By-product selling price (or disposal cost!)
> Product quality and whether or not it is satisfactory.
> Time taken to achieve first output.
> Time taken to produce at design rate.
> Output rate achieveable from new plant.
> Sales demand (made up of size of total market and share of market).
> Useful life of the plant.
> Obsolescence of the product.
>
>   etc.

To take a pessimistic view of all the risks would probably mean that one would never proceed. To take an optimistic view might be foolhardy.

It is true that in some cases unfavourable developments under some of these headings may be met by contingency plans such as marketing in an extra area or modifying the product. Also if one aspect in particular shows signs of giving trouble the effects may to some extent be mitigated by application of effort and resource. Nevertheless, for reasons already mentioned, it is preferable to be able to quantify the risk as far as possible.

The difficulty will be shortage of reliable data. Many of the headings mentioned will have no hard facts to substantiate them. Nevertheless the discipline of breaking the considerations into their several components is more likely to give a better picture than if the subject is treated globally.

The technique consists of taking each element in turn and analysing it in depth. Take raw material costs for example. The figure used in orthodox calculation is (say) £150 per tonne. If this is examined it may be concluded that there is no chance of this being less than £120 per tonne under any conceivable circumstances and one may judge that there is only a 10% chance of it lying in the range £120 to £140. On the other hand there may be a fair certainty (say 75%) it will be less than £160 and just an outside chance (say 2%) that it will be over £180. These may be subjective judgements but will presumably be backed by as much data as can be made available.

These figures can be represented graphically as in *Figure 31*.

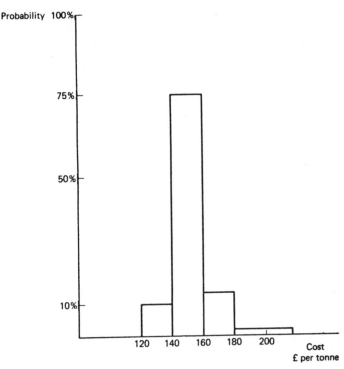

Figure 31. Probability of cost

A word of warning would be that the outside chances of wide variations should not be ignored. The example is hypothetical but is worth asking whether there is any chance of the raw material costing (say) £300 per tonne. All sorts of possibilities should be considered, e.g. world shortage, political changes, technical problems, etc.

The discipline which the preparation of this data requires is valuable to create a fuller understanding of the environment in which the project is proceeding. The full benefits however require that all the data on the variables having been gathered into forms similar to that shown in *Figure 31* should be combined. Since we are talking about probabilities there are an infinite number of ways in which the data can be collected. We are looking for the most probable combination of the data which ultimately will be expressed as a most probable rate of return. This step, because of the volume of the calculations, cannot reasonably be undertaken without a computer programme.

Over and above the expression of the most probable rate of return the other requirement is that the calculations shall reveal the sensitivity of that rate to changes in the variables. The postulation for example of a 10% change in capital required or a 10% change in sales forecast can result in a numerical comparison of these two changes and a real measure of the comparative sensitivity of the project to these variables. This type of analysis is particularly valuable in the comparison of two projects when there are only enough funds for one or to compare two processes which are significantly different in capital requirements and operating costs.

We have taken an example from the industrial scene but the method is equally applicable to the public sector. In this case however the criteria for the final judgement may well contain matters which are less amenable to arithmetical calculation.

Project Data
for a Chemical Plant

A typical chemical plant is chosen for this example of Project Data (see Chapter 3) partly because it is very much the author's personal experience and partly because it demonstrates the scope of different considerations which have to be built in to this stage. Of the headings which follow the first (Process) is mandatory and the root from which others grow. The other headings are not necessarily essential to any particular project but should be discarded only after deliberate consideration.The order is capable of being changed and indeed it will be found that there is so much interdependence between some aspects that preparation of a later stage of the data may fundamentally affect issues which had already been 'finally settled'.

Figure 32 will be found helpful as a pictorial summary of the position.

(1) Process

This requires basic statement of the end product of the action together with a description of the essential steps which are necessary to achieve it. In a chemical project the written process would be selected from any available choices of routes and alternative processes. This of itself might well have required a period of development or assessment of alternatives, reviews of patents and licenses, economic and reliability studies etc.

The purpose of writing down the process in a formal manner at this stage is to ensure that the essential chemistry is recognisable and correct.

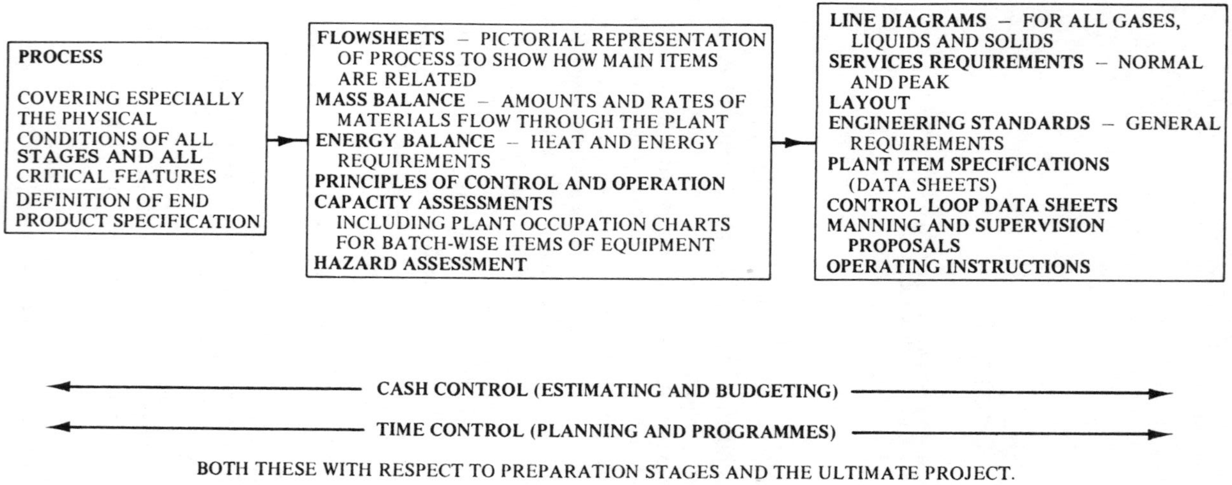

114

PROCESS

COVERING ESPECIALLY
THE PHYSICAL
CONDITIONS OF ALL
STAGES AND ALL
CRITICAL FEATURES
DEFINITION OF END
PRODUCT SPECIFICATION

FLOWSHEETS – PICTORIAL REPRESENTATION
OF PROCESS TO SHOW HOW MAIN ITEMS
ARE RELATED
MASS BALANCE – AMOUNTS AND RATES OF
MATERIALS FLOW THROUGH THE PLANT
ENERGY BALANCE – HEAT AND ENERGY
REQUIREMENTS
PRINCIPLES OF CONTROL AND OPERATION
CAPACITY ASSESSMENTS
INCLUDING PLANT OCCUPATION CHARTS
FOR BATCH-WISE ITEMS OF EQUIPMENT
HAZARD ASSESSMENT

LINE DIAGRAMS – FOR ALL GASES,
LIQUIDS AND SOLIDS
SERVICES REQUIREMENTS – NORMAL
AND PEAK
LAYOUT
ENGINEERING STANDARDS – GENERAL
REQUIREMENTS
PLANT ITEM SPECIFICATIONS
(DATA SHEETS)
CONTROL LOOP DATA SHEETS
MANNING AND SUPERVISION
PROPOSALS
OPERATING INSTRUCTIONS

CASH CONTROL (ESTIMATING AND BUDGETING)

TIME CONTROL (PLANNING AND PROGRAMMES)

BOTH THESE WITH RESPECT TO PREPARATION STAGES AND THE ULTIMATE PROJECT.

Figure 32. Project data

(2) Flowsheet

A typical flowsheet is shown in *Figure 6*. There are various interpretations as to how much detail ought to be shown on a flowsheet. There is probably no right answer but as a minimum it should show all the major plant items which are seen to have a separate identity. The physical conditions such as temperature and pressure, the materials of construction and the principal process lines should be identified so that the relationships between the items can be established. Some acknowledgement can be made to vertical height and gravity feeds although it should be accepted that these might be changed at the layout stage.

(3) Mass Balance

On the principle that what goes in must come out, a mass balance ensures that due account has been taken of all the component flow streams. It is surprising that the discipline of preparing a mass balance usually reveals some neglected aspects of the process; a by-product or effluent stream, a solids handling problem or a better appreciation of the water requirements. Later on the mass balance will be used for pump and pipe sizing, for definition of ancillary equipment and for confirming the sizes of all the components (*Figure 33*).

(4) Energy Balance (Heat Balance)

The most important aspect of the principle of the conservation of energy is to ascertain the heating and cooling requirements. Kinetic energy considerations are likely to be negligible but heats of reaction can be considerable. Otherwise the main consideration is to size the service facilities. A balance of this type ensures that significant considerations are not overlooked (*Figure 34*).

It is not strictly necessary to determine at this stage how the necessary energy will be provided or removed but that will be considered further under item 9 below.

(5) Principles of Operation and Control

At an early stage it is necessary to determine what principles will apply. This is necessary to get a basis for layout proposals and also to get the manning and supervision proposals into context.

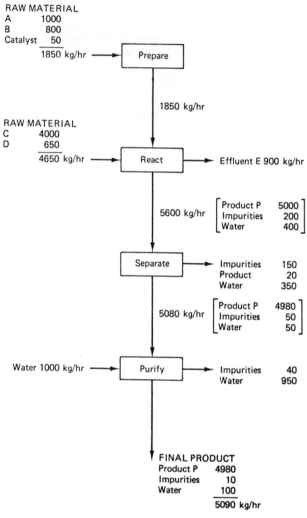

RAW MATERIAL
A   1000
B   800
Catalyst   50
            1850 kg/hr ⟶ Prepare

1850 kg/hr

RAW MATERIAL
C   4000
D   650
            4650 kg/hr ⟶ React ⟶ Effluent E 900 kg/hr

5600 kg/hr
$\begin{bmatrix} \text{Product P} & 5000 \\ \text{Impurities} & 200 \\ \text{Water} & 400 \end{bmatrix}$

Separate ⟶ Impurities 150
              Product   20
              Water   350

5080 kg/hr
$\begin{bmatrix} \text{Product P} & 4980 \\ \text{Impurities} & 50 \\ \text{Water} & 50 \end{bmatrix}$

Water 1000 kg/hr ⟶ Purify ⟶ Impurities   40
                              Water   950

FINAL PRODUCT
Product P   4980
Impurities   10
Water   100
             5090 kg/hr

Figure 33. Mass balance

The choice may range from extremely sophisticated proposals through to very elementary such as:

(a) On-line computer.
(b) Programmed operations.

(c) Centralised control room with remote pushbuttons.
(d) Dispersed control points with automatic operations.
(e) Manual operation based on indicated information.
(f) Minimum control mechanisms.

The choice will be influenced by the nature of the materials being handled, the degrees of criticality of the operation, the availability of effort to maintain sophisticated equipment and to operate any type of equipment provided, the custom on other adjacent plants, environmental considerations, economic factors, delivery problems, development issues etc.

| | HEAT INPUTS | | HEAT OUTPUTS | | |
| | | | By material | | |
	By material	Applied	Main stream	Side stream	Losses
Prepare to 60°C	1850 @ 20°C	1850 20 → 60°C + losses	1850 @ 60°C	Nil	Radiation
React at 90°C	4650 @ 20°C 1850 @ 60°C	4650 20 → 60°C 6500 60 → 90°C + losses − heat of reaction	5600 @ 90°C	900 @ 90°C	Radiation
Separate	5600 @ 90°C	Nil	5080 @ approx. 70°C	520 @ approx. 70°C	Radiation
Cooling to 20°C	5080 @ 70°C	Cooling 5080 70 → 20°C − losses	5080 @ 20°C	Nil	Radiation
Purify	5080 @ 20°C 1000 @ 10°C	Nil	5090 @ 10/20°C	990 @ 10/20°C	Negligible

Figure 34. Energy balance

Notes: 1. *Compare with Figure 33.*
2. *Quantities of materials in kg/hr.*
3. *Actual heat figures to be calculated and depend on specific heats of materials*

This tends to be an emotive subject and it is therefore valuable to get true expressions of requirements established and agreed by all interested parties at an early stage in the preparation of the data. Experience has shown that it is very easy for the initial intentions to drift away during the course of the design. If the drift is towards more sophistication it will tend to be more expensive in capital. Also there will be a tendency for unevenness through the plant which will lead to inefficiency of use, i.e. additional operational labour will be required but not fully employed.

Depending on the process to be employed it is worth emphasising that the reasons for wishing to have more sophisticated control systems are more likely to be quality control rather than labour economy. The tightness of the product specification is likely to be a dominant factor with restricted variations as the target.

(6) Capacity Assessments

It is most important that all the stages of the process are in balance. This is a function of the size of the item, the throughput rate and the time of occupation. These three are interrelated but any one might be

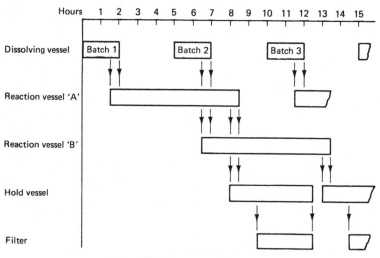

Figure 35. Plant occupation chart
Note: Because of limitations on the hold vessel in this case the true cycle time is 10 hr, i.e. maximum throughput is 1 batch every 5 hours

critical for a particular process or stage of a process. Their relationship will also determine the interstage storages which are required. Whether these are tanks or warehouses will vary with the material but it is necessary to consider normal operation, breakdown, recycling, start up and shut down conditions.

A batchwise plant can be dealt with satisfactorily by means of a plant occupation chart as shown in *Figure 35*.

(7) Hazard Assessment

However much subsequent design and construction teams may be expert in their particular line of business it is unreasonable to assume that they will know as much about the specific hazards of the project as those who have conceived the job. It is therefore necessary at an early stage of the activity to identify the hazards put them into perspective and comment on the means of reducing them. Hazards in this context include poisons, fire-risks, risks of suffocation, acids or strong alkalis, heat, risk of side reactions or lack of control, obnoxious conditions of any sort etc.

Getting hazards into perspective will not only lead to more effective action at subsequent stages in the project but will also contribute to a more environmentally acceptable outcome.

(8) Line diagrams

The object of a line diagram is to identify the needs for whatever piping and pipe-fittings are required on the job. They become necessary in chemical plants because of the complexities which are involved but it should be recognised that practice varies. For example more than one large plant has been built direct from a piping model which in turn was based on standards and specifications rather than on line diagrams.

The conventional line diagram however seeks to show every piece of pipe, valve, fitting or instrument required on the plant. It is primarily a mechanical engineering document because the other engineering disciplines will, of necessity, have to produce other drawings with their own requirements at the detailed design stage. They will however all spring from and be coordinated by the line diagram.

It should be possible to go through every process operation in fine detail on the line diagram to ensure that all operations whether manual or automatic produce the effect that is intended. It will also be possible to see that all the required information is available and is meaningful.

Start up and shut down operations can also be followed to ensure that they are acceptable. Recycling where necessary and cleaning out are also important.

It is often convenient to have a separate line diagram for each process unit and also separate diagrams for each engineering or chemical service. This however will vary with the complexity of the plant.

As a general rule not too much effort is made to get the geography of the plant correct on the line diagram because it is likely that some layout decisions will depend on the details developed on the line diagram. Nevertheless it is useful to secure reasonable representation.

(9) Services Requirements

For each service, e.g. water, steam, gas etc. it is necessary to establish the expected use. This must include the instantaneous peak usage rate (should all possible requirements be met is a question worth asking!) and also the pattern of usage. These are needed to establish source requirements, storage capacity and the capacity of the distribution system.

This has to be done in conjunction with the capacity assessments mentioned in paragraph 6 above. For example if cooling water is required it matters a great deal whether this is required 24 hours a day or for just 2 hours out of 24. In the second case the size of the cooling tower can perhaps be 80% of the size in the first case but the size of the hot well will be much bigger with the smaller tower.

This is the stage at which the value of energy is to be considered. If cooling is required from a high temperature the cooling medium itself may in turn be usable as a heating medium at a lower temperature. This may make a two-stage cooling operation more attractive overall than a single-stage system but of course it will depend on the demand for medium grade heat.

The choice of fuel will also have to be made if heat is required. Process considerations may influence the choice between steam, electric heating, direct gas or oil with or without heat transfer fluids. The energy input is likely to depend on the fuel economy at the time of decision, the other uses on the site, the certainty of continuity of supply, technical issues and so on as well as speculation into the forward energy position at the location of the plant.

Technical issues to be borne in mind will include the possibility of heat recovery at a useful temperature, the possibility of electricity generation and other matters which may be local to the particular problem.

Dependability and reliability are always important issues in the provision of services. Alternative supply systems should be considered especially if the processes to be achieved are continuously operated.

The range of services under this heading should be as comprehensive as possible. There may, for example, be two or more grades of water available or required. There may be several different steam pressures required and a rationalisation may be valuable. It is also likely that compressed air is wanted for more than one purpose, e.g. instruments, tools, material transfer and breathing air. To what extent is a rationalisation possible or desirable? (Note: there must never be any risk of contamination of a breathing air supply and delicate instruments should also be treated with respect.)

The consideration should also include fire-fighting services, telephones and intercommunication systems because all will be easier to incorporate if considered at the right time.

(10) Layout

The layout of a chemical plant can be more complex than the layout of equipment in many other industries. This is especially true if the equipment is on more than one floor or if, as in a refinery type of structure, there are many different operating levels with interconnections for access and maintenance.

Without considering how best to tackle the problem of achieving the layout, because different engineers have their own pet theories, we should consider what we want to achieve.

We are attempting to facilitate later stages of design without too much alteration to the concept of the scheme. At this stage we therefore need to establish the locations of all the major and most of the minor items of equipment. The locations need not be absolutely precise because some sizes and positions of supports may not yet be known. The degree of precision however will vary from item to item. For example: if a piece of equipment weighs five tonnes it is likely that the floor which supports it will be specially designed so the structural designer will want to locate it precisely at an early stage. If it is subsequently moved by a couple of metres this could fundamentally affect the structural design. On the other hand a control panel weighing a few hundred kilograms could perhaps be moved a lot more or even introduced new without major problems.

The preferred action then is to locate the items and fix some of the more crucial dimensions—particularly those which affect structural design. The action must meet certain requirements, i.e.

(a) It must be comprehensive. Every item must be allowed for. This can be verified by means of the line diagrams.
(b) It must take account of material movement and the availability of services and facilities.
(c) It must make operational sense. Regard must be given to movements of operators and supervisors. It is also necessary to consider what information they need to undertake their work and where that information is available.
(d) Maintenance must be considered. What has to be dismantled occasionally or frequently? What has to be adjusted? Where do the maintenance workers do their work etc?
(e) It must take account of the hazard assessment.
(f) It must be economical of space without being cramped and without inhibiting alteration or addition if this is seen as possible.

Some layout studies are achieved by the use of models—a distinct aid if there are three dimensional problems. Others are achieved entirely on paper. Whichever method is adopted there is great advantage in committing the agreed (final?) layout to paper and keeping a record whatever may subsequently be decided. This is not so much for personal vindication as to keep track of the intentions at the time agreement is reached.

(11)   Engineering standards

It may be sufficient to express the desire that all the design work shall follow the appropriate British Standard specification or Code of Practice. It may be however that American or Continental practices are to be preferred for some items. Alternatively the company may have its own or some special standards which it wishes to follow. Yet again some aspects of the project may be so special or novel as to warrant a specification or standard being written for this job.

This is the time to clear the air if there are any special requirements. If not it may be assumed that subsequent work will be done in accordance with normal practice at the office or site where the work is carried out. This may be perfectly acceptable but the option should be considered.

(12)   Plant Item Specifications

The use of data sheets to express a user's requirements is a widespread practice although the precise form of the sheet varies considerably. A typical sheet is shown in *Figure 36* and this covers the essential requirements.

VESSEL DATA SHEET   PROJECT:
                                         LOCATION:
TITLE: ITEM NUMBER:

CAPACITY: MAX: WORKING:
CRITICAL DIMENSIONS: (SKETCH IF NECESSARY)
CONTENTS:
TEMPERATURE: DESIGN: OPERATING:
PRESSURE: DESIGN (RELIEF VALVE):   OPERATING:
MATERIAL SPECIFICATION:
     CORROSION ALLOWANCE: RADIOGRAPHY: STRESS RELIEF:
     DESIGN CODE OR SPECIAL REQUIREMENTS:

CONTENTS SPECIFIC GRAVITY: VISCOSITY:
AGITATOR: TYPE: REF. NO:
COIL: AREA: BORE:
      FIXING: REF. NO:
END DETAILS: TOP:
             BOTTOM:
SPECIAL FITTINGS:

     BRANCHES:

                              (SEPARATE SHEET IF REQUIRED)

VESSEL LOCATION:
MEANS OF SUPPORT:
WILL VESSEL BE BOILING (VIBRATING) AT ALL?:
INSULATION:
PAINTING:
SPECIAL REQUIREMENTS OF ANY SORT:

REFERENCE DRAWINGS OR DATA:

DATE OF PREPARATION:
ORIGINAL OR AMENDMENT:
PREPARED BY:
CHECKED BY:

Figure 36. Plant item data sheet

Appendix 3

(13) Control Loop Data Sheet

Not only is it necessary to specify a particular function which may be required but also it is necessary to take into account some details of that function. For instance: the ability to over-ride, the sensitivity of

---

CONTROL LOOP DATA SHEET   PROJECT:
PRESSURE   LOCATION:
                              PLANT ITEM NUMBER:
                                      LOOP NO:

GENERAL STATEMENT OF REQUIREMENTS:

PRESSURE POINT LOCATION:

INSTRUMENT LOCATION(S):

PRESSURE TO BE:– RECORDED/INDICATED/CONTROLLED.

FLUID COMPOSITION:

PRESSURE: MAXIMUM: OPERATING:
TEMPERATURE: MAXIMUM: OPERATING:
FREEZING POINT: BOILING POINT:
SPECIFIC GRAVITY: CONDITIONS:

MATERIALS OF CONSTRUCTION COMPATIBLE WITH FLUID.

IS MERCURY CONTACT PERMITTED?:
WHAT PURGE ARRANGEMENTS ARE EXPECTED?:

IS PLANT CONNECTION FLANGED OR SCREWED?:
            WHAT SIZE?:

WHAT IS FIRST ISOLATION VALVE?:

REFERENCE DRAWINGS OR DATA:

SPECIAL REQUIREMENTS OF ANY SORT:

DATE OF PREPARATION:
ORIGINAL OR AMENDMENT:
PREPARED BY:
CHECKED BY:

Figure 37. Control loop data sheet

response, the precise action required, shut down and start up actions, failing to safety etc. The discipline of preparing a control loop data sheet is intended to provide answers to those questions. Again practice varies but an example is shown in *Figure 37*.

(14) Manning and Supervision Proposals

This should follow the Principles of Operation and Control discussed in Section 5 above. There will have been a crystallisation of ideas in working through line diagrams and layout and this phase should not pass without a check to ensure that the original ideas on manning and supervision are still valid and capable of being implemented. In any case consideration of numbers and duties will affect the final location of control points, alarm and intercommunication design, provision of amenities, car parking etc. It will also of course affect the operating cost of the project and it is likely that a review of the project economy will be undertaken as the project data are being finalised.

(15) Operating Instructions

Many plants are built without operating instructions being written down. Then managers wonder why they need too many men to run the plant or why the plants don't work properly. Failure to write down operating instructions usually denotes failure to think through the requirements in detail. It is this failure to consider the job in depth which is more serious than the non-arrival of bits of paper.

Operating instructions are tedious to write and yet the discipline is valuable. The problems ought to have been considered and overcome at the preparation of the line diagrams or the approval of the layouts. Writing down the requirements is therefore a logical step at this stage. It also ensures that the data assembled does not conflict with itself in any way. If it does not the path of subsequent design will be that much smoother.

Project Data
for a Swimming Bath

Following the discussion of the need for Project Data in Chapter 3 this example is chosen to show how the principles can be applied to a project in the Public Sector. It is emphasised that there is no absolute rule as to what should or should not be included in data of this sort. The guideline must always be that data are collected in this way for the clarification of the objectives and to lay down requirements for the detailed design activity which must follow.

Equally there is no great significance in the particular order chosen for the following sections. Broadly speaking the sequence for any particular project will be determined by the starting point for that project and a logical progression through the headings which are relevant. In any case, as noted in Chapter 3 some backtracking during the preparation of this data is to be expected.

(1) Basic Statement of Main Benefits Required

The bath complex might include one, two or more pools depending on the provision that is to be made for separate diving, children, water polo etc. In each case principal dimensions must be laid down. Consideration should be given as to whether Olympic standards are to be met or not.

The changing accommodation must also be reviewed and numbers established. This will include considerations of cubicles, showers, footbaths, toilets etc. Separate numbers would be given for men and women but also policies on interchangeability and segregation. Sauna baths also might be provided.

Provision must be made for spectators and possibly also press and television facilities. This would cover the sizes and requirements of the viewing gallery/galleries and waiting rooms. Judges for competitive events must also be considered and provision made.

(2) Assumptions and Calculations

In a project of this type it is advisable to ensure that there is an adequate record of the data and assumptions which have gone into the determination of size. One hopes that changes of a radical nature will not need to be made at a late stage but a record will be invaluable if changes are contemplated.

For a swimming bath the issues covered in this section would probably include:

(a) The population in the area now and predicted for the future.
(b) The location of other nearby facilities of a similar type. How far away are they and how good are they?
(c) What information is available on national trends of popularity? Are the trends influenced by legislation, financial aid, new records etc?
(d) What is the success record of other facilities built elsewhere and what factors influenced their success?
(e) Are any local factors likely to influence the predictions, e.g. attitudes of local authority, local politics, factors affecting the social habits of the population etc.
(f) Is there any risk of additional competition from other leisure facilities? (These could be either public or private ownership.)

From these assumptions, calculations will be made to show the capacity of the facilities at all stages.

(3) Auxiliary Facilities

This would cover the possibility of using the facilities for alternative purposes on occasions. Such uses might include a ballroom, an ice-rink, a bingo hall, exhibition facilities etc.

There is also the question of any adjacent land or buildings having an impact on the facilities included in the project. This might be the case if there were an adjacent playing field demanding changing accommodation or even a bowling green demanding storage space for rollers and mowers.

All these will affect the auxiliary facilities in the building which would probably include such items as a restaurant or snack bar, first aid room and facilities, public telephones etc.

A special feature which applies to a swimming bath complex would be to consider the needs of the disabled and what special facilities should be provided.

(4) Operating Policies

The way in which the baths are to be operated and managed will have an important effect not only on the facilities to be provided but on the economics of operation. Early decisions on such policy matters will be helpful although it is to be expected that some flexibility will be required.

Consideration should be given to such matters as:

(a) Charging for admission. Not only prices but also how the money is to be collected. If by tickets should they be from vending machines or kiosk? Are spectators to be charged? How should we cope with parties or season ticket holders? etc.

(b) Internal control. Is a time limit for swimmers to be introduced? How will it be managed? Are different baths segregated? Are 'spectators' prevented from swimming? What supervision and supervisors' accommodation is required? What arrangements will be made for lost property? etc.

(c) Lighting. Is advantage to be taken of daylight when available? To what extent is artificial light essential?

(d) Cleaning. Are the cleaning requirements understood and catered for? What equipment? What staff? What accommodation for equipment and staff? How is the water replaced and treated?

(e) Maintenance. Who will do it? What facilities and accommodation are required? What standby pumps/filters/lights etc. are required?

(f) Offices and administration. What accommodation is required? Is there a need for meeting rooms of any sort. What facilities? Telephone requirements?

(5) Transport

When it is established how many people may be expected to work in or use the facility the transport situation should be reviewed. Will those people arrive on foot, by car, by bus, by coach? Adequate facilities by way of roads, parking areas, unloading areas etc. must be provided.

Maintenance requirements will also need transport for special equipment or people, spare parts and stores.

If there is a significant catering facility there will also be the question of transport to the kitchens and store rooms of that facility.

(6) Layout

It may not be necessary in this particular case to determine the layout of the facility as this will demand a considerable amount of detailed investigation to optimise from all the requirements. It would be as well however to identify specially important features. Some of them may be crucial to the validity of the assumptions in the data and should therefore be singled out for special mention so that the requirements are not neglected.

(7) Heat Balance

All temperature levels should be determined. This applies to the water temperature and the air temperature in various parts of the building. Humidifying or dehumidifying requirements should also be stated. The data in this section will permit insulation standards to be established and the overall energy requirements to be assessed.

(8) Services Requirements

From the consideration of energy it should be possible to determine the preferred method of heating. To some extent the choice will be on grounds of economy but operational and maintenance considerations must be included. This is the appropriate time to consider the contribution of solar energy in addition to the conventional fuels of electricity, oil, gas or coal.

(9) Standards

Wide choices are available and much will be left to the detailed design but some general guidelines are appropriate. The range between utility and de luxe is wide and assembly of the data so far will have been based on some view of the finished job.

Equally it is possible to vary the standards quite significantly in different parts of the building. To quote an extreme example no one would expect the maintenance stores to be comparable to the manager's office or the spectators' gallery.

Standards also apply to external appearance and again a wide range of possibilities exist. Guidelines in line with the economic calculations need to be laid down.

(10) Safety

Consideration should be given to safety for two main reasons. The intention is obviously to provide a popular facility which will be enjoyed simultaneously by many people. Whenever people are gathered together risks of fire or other emergency must be considered. This will affect passageways, doors, staircases, fire escapes, ventilation etc.

In addition there are obviously specific risks in a swimming bath. The risk of drowning by a swimmer (or a spectator!) must be recognised but also there may be equipment which itself introduces an element of risk.

It is appropriate at this comparatively early stage of the design activity to review the situation. This would be aimed at identifying the risks and specifying safeguards and precautions so that the risks are at a satisfactory level.

Further Reading

General

Control of Engineering Projects. Ed S. H. Wearne. London; Edward Arnold, 1974
Up the Organisation. Robert Townsend. London; Michael Joseph, 1970.
Problems and Efficiency in the Management of Engineering Projects. Symposium at the University of Manchester College of Science and Technology. April 1966.
Successful Project Management. W. J. Taylor and T. F. Watling. London; Business Books Ltd, 1973.
Practical Project Management. W. J. Taylor and T. F. Watling. London; Business Books Ltd, 1973.
The Basic Arts of Management. W. J. Taylor and T. F. Watling. London; Business Books Ltd, 1972.
Project Management. Denis L. Lock. London; Gower Press, 1968
The Commercial Management of Engineering Contracts. Peter Scott. London; Gower Press, 1974.
Successful Engineering Management. Tyler G. Hicks. New York and London; McGraw-Hill, 1966.

Chapter 1

The Public Client and the Construction Industries. N.E.D.O. London; H.M.S.O., 1975.
Before you Build. N.E.D.O. London; H.M.S.O., 1974.
'Project management, Swedish style'. Per Jonason. *Harvard Business Review*. Nov. 1971. pp 104–109.
The LKAB Method—Flexible Project Organisation. Per Jonason and Bert Lindkvist. LKAB, 1973.

Chapter 2

Appraisal and Control of Project Costs. R. Pilcher. New York and London; McGraw-Hill, 1973.

'Investment decisions—evaluation techniques in perspective'. D. H. Allen. *The Chemical Engineer,* Jan. 1975. pp 42–45.

Safety and Health at Work. Report of the Robens Committee. London; H.M.S.O., 1972.

Annual Reports. H. M. Chief Inspector of Factories. H.M.S.O.

Health and Safety at Work Act 1974. London; H.M.S.O., 1974.

Chapter 3

Reviewing the Management Structure. London; British Institute of Management, 1972.

Chapter 4

Civil Engineering Bills of Quantities. N. M. L. Barnes and P. A. Thompson. *C.I.R.I.A. report 34.* Sept. 1971.

Civil Engineering Standard Method of Measurement. London; Institution of Civil Engineers, 1976.

Chapter 5

Critical Path Analysis. D. W. Long. London; E.U.P. Teach Yourself Books, 1970.

Chapter 6

Management Training for Engineers. Barry T. Turner. London; Business Books Ltd, 1969.

Work and the Nature of Man. Frederick Herzberg. New York; World Publishing Co., 1966.

Behavioural Science in Management. Saul W. Gellerman. London; Penguin, 1974.

Chapter 7

Total Technology. Science Research Council, July 1973.

Getting buildings designed and built – can we afford today's ways. Institute of Building and Property Services Agency One Day Conference. April 1975. Institute of Building.

Time Cost and Architecture. George T. Heery. New York and London; McGraw-Hill, 1975.

Index